FROGS

OF THE WORLD

FROGS
OF THE WORLD

A GUIDE TO
EVERY FAMILY

Mark O'Shea and
Simon Maddock

PRINCETON UNIVERSITY PRESS
PRINCETON AND OXFORD

Published in 2024 by Princeton University Press
41 William Street, Princeton, New Jersey 08540
99 Banbury Road, Oxford OX2 6JX
press.princeton.edu

Conceived, designed, and produced by
The Bright Press
an imprint of The Quarto Group
1 Triptych Place, London, SE1 9SH, United Kingdom
T (0) 20 7700 6700
www.quarto.com

Library of Congress Control Number 2023940400
ISBN: 978-0-691-24830-1
Ebook ISBN: 978-0-691-24831-8
British Library Cataloging-in-Publication Data is available

Publisher James Evans
Editorial Director Anna Southgate
Managing Editor Jacqui Sayers
Art Director and Cover Design James Lawrence
Senior Editors Joanna Bentley and Dee Costello
Project Manager Sara Harper
Design Wayne Blades
Picture Research Susannah Jayes
Illustrations John Woodcock

Cover and prelim photos: Front cover (clockwise from top left):
Shutterstock / Petlin Dmitry; Shutterstock / J. D. Carballo;
Shutterstock / Eric Isselee; Shutterstock / Krisda Ponchaipulltawee;
Shutterstock / Rosa Jay; Shutterstock / Dirk Ercken; Shutterstock /
Rosa Jay; Shutterstock / Luis Louro; Dreamstime / Farinoza;
Shutterstock / Rosa Jay; Shutterstock / Natthawut Ngoensanthia;
Wikimedia Commons / Stephen Zozaya; Shutterstock / Pumidol.
Spine: iStock / Reptiles4All. **Back cover**: Shutterstock / Aastels.
Page 2: Shutterstock / John Copland. **Page 5**: Shutterstock /
Chase D'animulls.

Printed in Malaysia

10 9 8 7 6 5 4 3 2 1

CONTENTS

INTRODUCTION

Frogs are amphibians, occurring on every continent except Antarctica. They are characterized by being ectothermic (traditionally referred to as "cold-blooded") and possessing four limbs but lacking a tail. In many temperate regions, frogs are considered ubiquitous with watercourses, where large raucous choruses can be heard during their breeding season. However, many frog species, especially in tropical regions, do not require water to breed and do not congregate in such large numbers. Found in water, on land, in the trees, and underground, frogs occupy many different ecological niches.

From bright reds to vibrant greens, frogs can be fantastically colored and even multicolored. Many species are able to change color. Their eyes are even more

ABOVE | The arboreal Red-eyed Tree Frog (*Agalychnis callidryas*) from Central America.

fascinating than the color of the skin, with various-shaped pupils and brilliantly patterned irises. They have one of the most dramatic life cycles of any vertebrate, completely changing their anatomy and physiology between the tadpole stage and the adult frog. Frogs can be among the smallest vertebrates in the world, reaching sizes of just ¼ in (8 mm), and growing to sizes of 12½ in (32 cm) and weighing up to 7½ lb (3.3 kg).

In environments where they occur, frogs are critical for ecosystems to function properly. Yet, worryingly, they are the most threatened land vertebrate group. Not only are they important ecologically but also from a human-centered perspective. Frogs primarily eat what many would consider to be pest invertebrates, and in some areas where frogs have declined, malaria prevalence has increased, creating a major medical and financial burden on those regions. People in some countries consume frogs, which form a major staple of their diets. Additionally, throughout history, frogs have had other uses; for example, indigenous peoples in South America used to coat their hunting darts with frog poisons to quickly immobilize prey.

Frogs have been intrinsically linked with humans since the dawn of our existence, appearing across the world in stories and texts on all continents where frogs occur. Some myths and beliefs were good and some bad; for example, killing a frog could lead to either flooding (New Zealand's Māori people) or drought (South African tribes). The early Aztecs, ancient Egyptians, and medieval Europeans had goddesses depicted as frogs in some form. In Britain, according to folklore, carrying a desiccated frog around your neck was believed to prevent epileptic seizures.

BELOW | A bas-relief of Pharaoh Seti I making an offering to the frog-goddess Heqet, in the Great Temple at Abydos, Egypt.

EVOLUTION AND TAXONOMY

Extant (opposite of extinct) species of frogs and toads account for nearly 90 percent of all amphibian species alive today, numbering over 7,600 individual species, with new species being described by scientists every year.

Amphibians belong to the class Amphibia and the frogs and toads to the order Anura. In classification terms, a true grouping is one that includes all species that evolved from a common ancestor. These are known as monophyletic groups. Frogs and toads are common names, and the name "toad" appears multiple times in different parts of the anuran tree of life (that is to say, it does not form a monophyletic group). As toads do not form a natural group, unless specifically talking about species that are considered true toads (family Bufonidae), we will refer to all members of the Anura as frogs. When a word ends with "-idae" it indicates that the unit is a family. If we are referring to a smaller grouping of species, we refer to this as a subfamily, with the specific grouping ending in "-inae."

At the time of writing, the frogs are placed into 56 families. However, taxonomy is a dynamic field and scientists regularly make taxonomic changes based on new and emerging research, so new families, genera, and species are continually being described or several species may be merged into one (known as synonymizing). Our taxonomy follows that of Amphibian Species of the World (https://amphibiansoftheworld.amnh.org/).

When common names are available we have used them and we have coined appropriate names where none are available, but we will use scientific names throughout this book to provide clarity.

EARLY FROGS

Extant amphibians form a monophyletic group known as the Lissamphibia, which contains the frogs and the two remaining orders, the salamanders + newts (order Caudata) and the caecilians (order Gymnophiona). The Amphibia have a long but not fully understood evolutionary history, with several extinct groups contained within the Temnospondyli, which would have dominated the Carboniferous period. Based on fossil evidence, the Anura and Caudata, a group known as the Batrachia, likely diversified from their ancestor, the now-extinct *Gerobatrachus hottoni*, which is thought to have first appeared around 290 million years ago (MYA).

Frogs started appearing approximately 250 MYA, with *Triadobatrachus massinoti* thought to be the ancestor of all modern frogs, looking very frog-like in its anatomical features. By 200 MYA, frogs were almost identical in body form to modern frogs. The fossil of the small *Vieraella herbsti*, measuring just $1\frac{1}{4}$ in (3 cm), shared almost identical features to those seen in today's frogs, including long, strong hindlimbs for jumping (discussed further in the Archaeobatrachia section).

PLATE TECTONICS

When the first frogs appeared in the fossil record 250 MYA, the world would have looked very different to how it does today. A single supercontinent known as Pangea existed. This would have enabled relatively free movement across all landmasses on the earth at that time. At approximately 200 MYA the two other major supercontinents that existed following the breakup of Pangea started to fragment themselves.

Gondwana consisted of the landmasses that now comprise South America, Africa, India, Madagascar, Seychelles, Antarctica, Australia, and New Zealand. Laurasia contained the modern landmasses of Eurasia and North America.

When we look at evolutionary relationships, it is clear that families that are each other's closest relatives (especially ones that diversified early in the history of frogs) are scattered across the globe. It is therefore fair to assume that these frogs, or their ancestors, had a wide distribution across Gondwana and Laurasia before they started separating. The breaking up of these supercontinents would have provided new challenges and opportunities—with new climates to adapt to, the formation of mountain ranges causing isolation,

ABOVE | Representation of *Triadobatrachus massinotii*, based on an Early Triassic fossil found in modern-day Madagascar.

RIGHT | Representation of *Gerobatrachus hottoni* based on an Early Permian fossil found in modern day Texas, USA.

and novel ecosystems forming, each of which would have allowed frogs to speciate into the fantastic forms that we see today.

Frogs are poor dispersers over marine barriers, compared to other organisms. Because they have porous skin, which they rely on to breathe, salt water can cause serious problems, even death. While some overseas dispersals have been reported, this is limited, and this is why very few frogs inhabit volcanically formed islands.

EVOLUTIONARY RELATIONSHIPS

Phylogenetic trees (trees henceforth) are used to understand evolutionary relationships between organisms. Traditionally, trees were constructed based on morphological characteristics, but since the rise of molecular techniques, these data have been used much more frequently.

Trees depict how taxa are related to one another, and these can be as large (depicting the entire tree of life) or small (depicting geographic differences within a single species) as desired.

When we have a split in a tree (in other words, where a single branch splits into more than two branches), this is known as a node and represents the common ancestor of everything that evolved after that split. We discussed monophyletic groups earlier, and these splits (nodes) in the tree represent monophyletic relationships, with everything that comes after the split being monophyletic. As trees can be as large or small as desired, so can monophyletic groups. These monophyletic groups can also be referred to as clades.

Trees are split into a series of sister relationships. The term "sister" can refer to sister taxa or relationships observed between taxa on trees. For example, if we refer to the phylogenetic tree opposite, we can say that the Ascaphidae and Leiopelmatidae are sister families. We can also say that a group containing the Ascaphidae and Leiopelmatidae is sister to all other extant frog families.

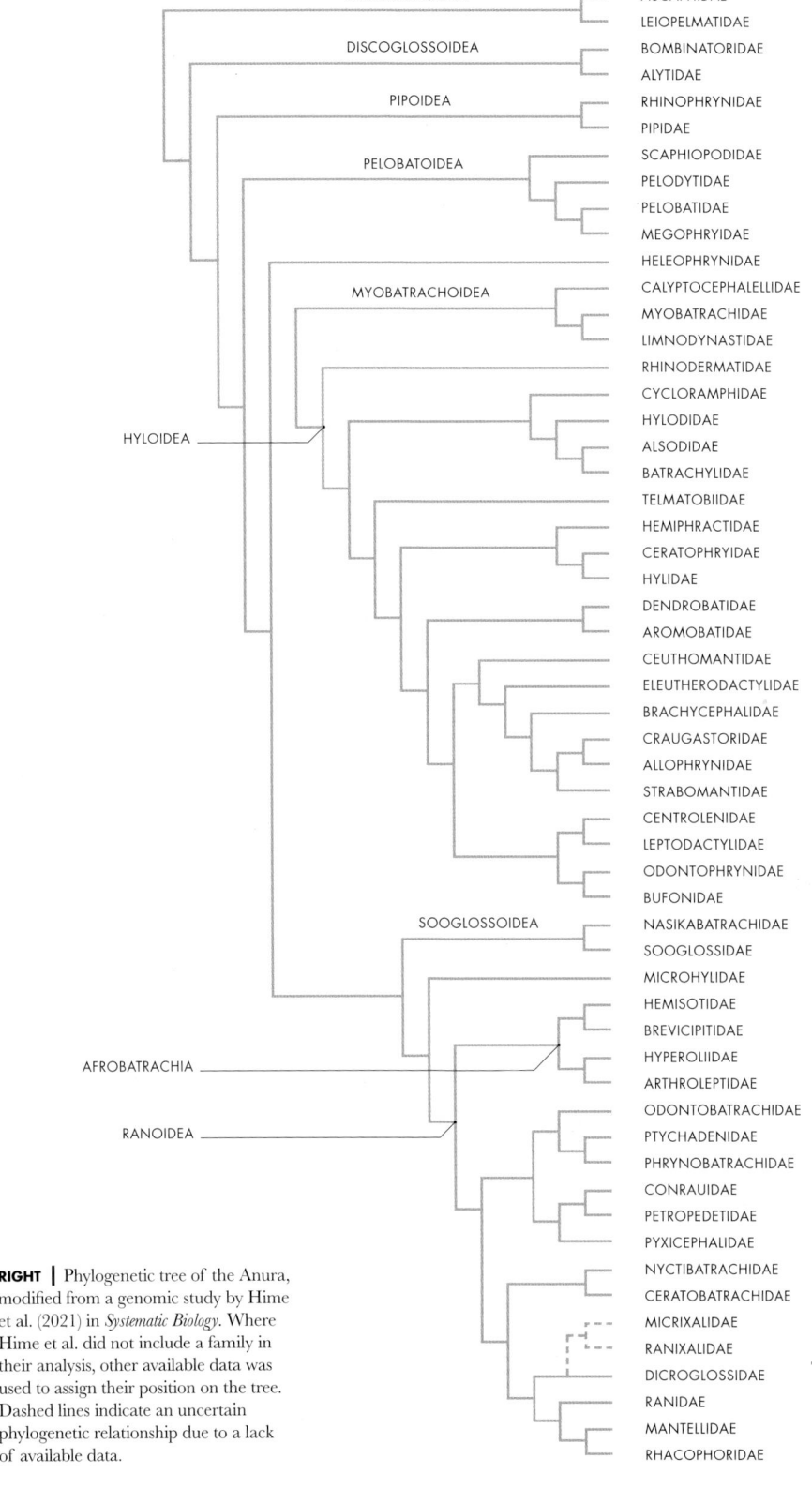

LEIOPELMATOIDEA — ASCAPHIDAE
LEIOPELMATIDAE

DISCOGLOSSOIDEA — BOMBINATORIDAE
ALYTIDAE

PIPOIDEA — RHINOPHRYNIDAE
PIPIDAE

PELOBATOIDEA — SCAPHIOPODIDAE
PELODYTIDAE
PELOBATIDAE
MEGOPHRYIDAE

HELEOPHRYNIDAE

MYOBATRACHOIDEA — CALYPTOCEPHALELLIDAE
MYOBATRACHIDAE
LIMNODYNASTIDAE

RHINODERMATIDAE
CYCLORAMPHIDAE
HYLODIDAE
ALSODIDAE
BATRACHYLIDAE

HYLOIDEA — TELMATOBIIDAE
HEMIPHRACTIDAE
CERATOPHRYIDAE
HYLIDAE
DENDROBATIDAE
AROMOBATIDAE
CEUTHOMANTIDAE
ELEUTHERODACTYLIDAE
BRACHYCEPHALIDAE
CRAUGASTORIDAE
ALLOPHRYNIDAE
STRABOMANTIDAE
CENTROLENIDAE
LEPTODACTYLIDAE
ODONTOPHRYNIDAE
BUFONIDAE

SOOGLOSSOIDEA — NASIKABATRACHIDAE
SOOGLOSSIDAE

MICROHYLIDAE
HEMISOTIDAE
BREVICIPITIDAE
AFROBATRACHIA — HYPEROLIIDAE
ARTHROLEPTIDAE

ODONTOBATRACHIDAE
PTYCHADENIDAE
PHRYNOBATRACHIDAE
RANOIDEA — CONRAUIDAE
PETROPEDETIDAE
PYXICEPHALIDAE
NYCTIBATRACHIDAE
CERATOBATRACHIDAE
MICRIXALIDAE
RANIXALIDAE
DICROGLOSSIDAE
RANIDAE
MANTELLIDAE
RHACOPHORIDAE

RIGHT | Phylogenetic tree of the Anura, modified from a genomic study by Hime et al. (2021) in *Systematic Biology*. Where Hime et al. did not include a family in their analysis, other available data was used to assign their position on the tree. Dashed lines indicate an uncertain phylogenetic relationship due to a lack of available data.

ANATOMY AND PHYSIOLOGY

The word amphibian roughly translates as "two" (Greek *amphi*) "life" (Greek *bios*) stages, in reference to most species having: (1) a larval stage, and (2) an adult stage. The larval stage in frogs is known as a tadpole, and although the two other extant orders of amphibian do have a larval stage, theirs is a much less dramatic ontogenetic change (change in body form between larva and adult) than that of the Anura. In fact, no other terrestrial vertebrate has such a dramatic shift in body plan between larva and adult as the frogs. Although the amphibians get their name from their two life stages, many species are direct-developing (especially in tropical environments), meaning that the young hatch out as miniature replicas of the adults, much like humans.

EGGS

Frog eggs have no shell, being much more gelatinous than those of birds, reptiles, and monotreme animals. Although not all frogs lay eggs in water, the adaptation to having shell-less eggs is due to them generally being deposited in an aquatic environment, where shelled eggs typically would not be able to survive. The jellylike substance surrounding the embryos provides the protection and nutrition required by the developing embryo.

TADPOLES

Tadpoles have a completely different body plan to adult frogs. At the early stages of development, they lack limbs but possess a tail. They contain no bones, instead having cartilage. They have keratinous jaw sheaths and toothlike denticles instead of teeth. Most have internal gills and a branchial basket, which supports the gills. The gills are not visible on tadpoles but instead small openings (spiracles) are present just behind the head. Spiracles are the exit holes for water that has passed over the gills after being taken in through the mouth.

The keratinized jaw structures in frogs are highly variable and reflect adaptions for feeding. Most tadpoles will feed on plant matter or detritus but some species are carnivorous. Mexican Spadefoot Toads (*Spea multiplicata*) are amazingly adaptable with two tadpole

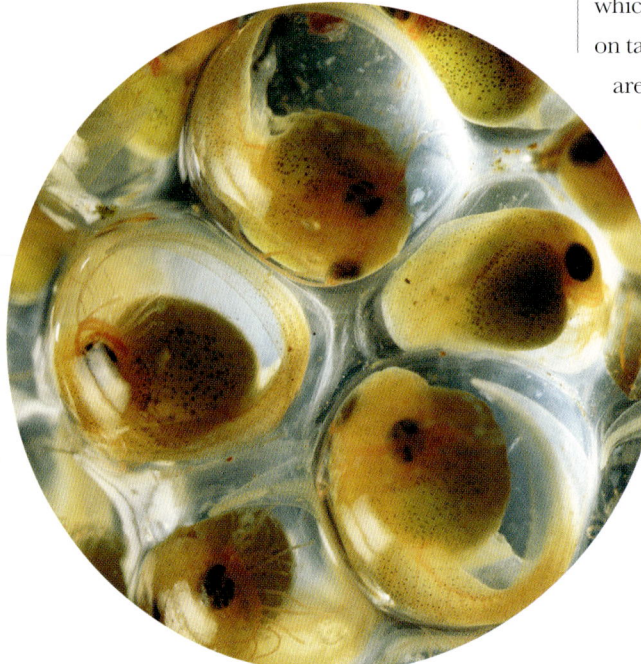

ecomorphs. Both the omnivorous ecomorph and the carnivorous ecomorph can be found in individuals from the same egg clutch.

The process of turning into an adult frog is known as metamorphosis and requires dramatic physiological and morphological changes. The major nonhormonal changes include: the shortening of the gut; the development of an acidic stomach; the appearance of jaws, a tongue, and teeth in some species; the development of lungs; the formation of moveable eyelids; the development of limbs; the ossification of cartilage into bone; the formation of skin glands and complex epithelium; and the reabsorption of the tail. Some tadpoles will become huge before undergoing metamorphosis, often growing to greater sizes than adults. Tadpoles of the Paradoxical Frog (*Pseudis paradoxa*) can grow three to four times the size of adult frogs.

Some frogs have very brief larval stages lasting only eight days whereas others have extended larval periods. Tadpoles of the Coastal Tailed Frog (*Ascaphus truei*) can take up to four years to metamorphose due to living in cold, fast-flowing streams, where physiological processes are reduced due to the cold conditions.

ABOVE | This American Bullfrog (*Aquarana catesbeiana*) tadpole has started to grow hindlimbs in a process known as metamorphosis. Hindlimbs always form before forelimbs appear.

BELOW | Body structure of a typical tadpole viewed from the side.

TADPOLE

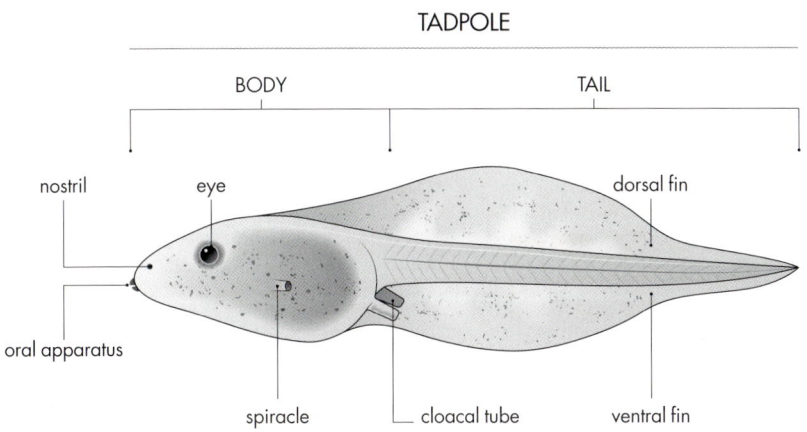

BODY   TAIL

nostril   eye   dorsal fin

oral apparatus

spiracle   cloacal tube   ventral fin

ADULTS

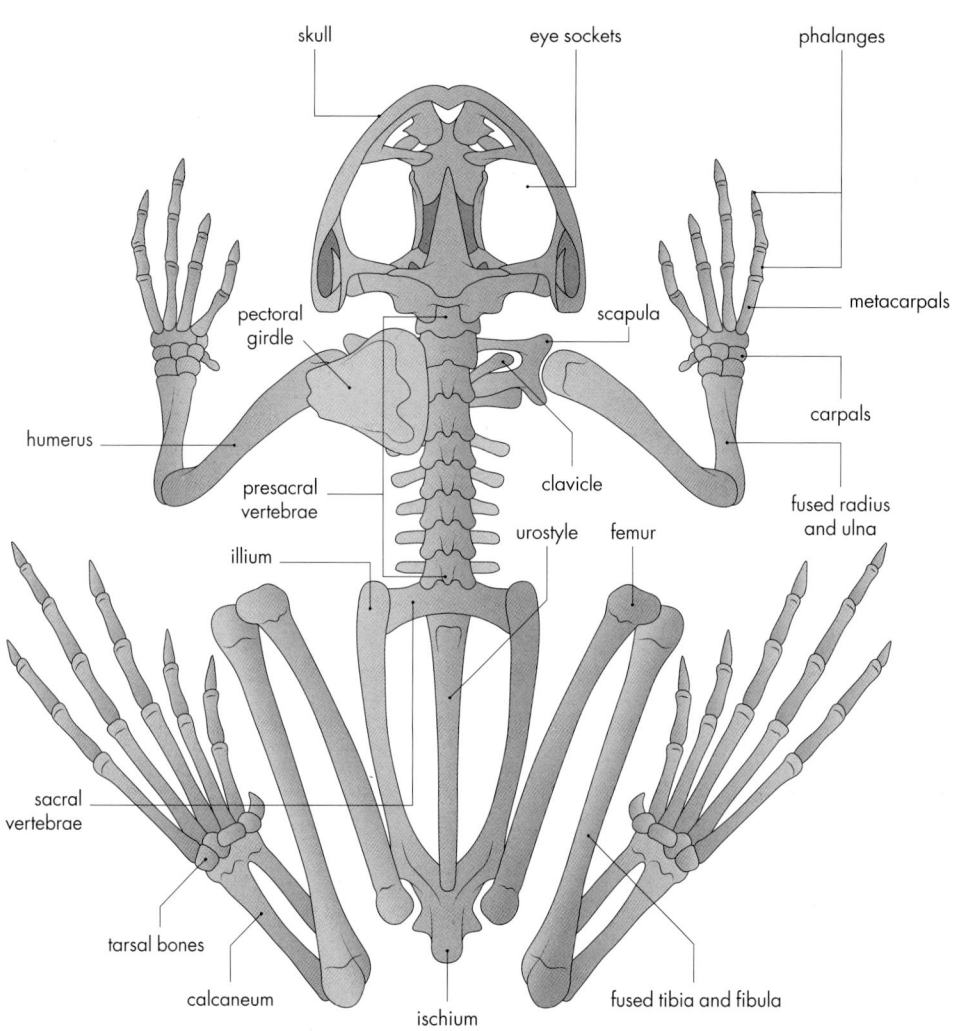

skull

eye sockets

phalanges

metacarpals

pectoral girdle

scapula

carpals

humerus

clavicle

fused radius and ulna

presacral vertebrae

urostyle

femur

illium

sacral vertebrae

tarsal bones

calcaneum

ischium

fused tibia and fibula

SKELETONS

The skeletons of adult frogs share a lot of the same basic body plan as other tetrapod vertebrates, consisting of a skull with a lower mandible, a vertebral column, a pectoral girdle attached to the forelimbs, and a pelvic girdle attached to the hindlimbs. Most modern frogs, however, have no ribs and no anatomical tail.

Frog skulls are very reduced and lack many of the bones present in other amphibians and vertebrates.

ABOVE | Skeleton of a typical frog viewed from above.

RIGHT | Budgett's Frogs (*Lepidobatrachus laevis*) have an exceptionally large mouth and head. They remain motionless in soft mud or water with just their eyes and nostrils protruding above the surface waiting for an unsuspecting meal to come close.

This evolutionary loss of bone makes the head lightweight and enables the frog to move through its environment more efficiently. The big openings on the top of the skull also house the relatively large eyes of anurans. Some species have, however, adapted to have more robust skulls, a process known as hyperossification, and these are associated with differing life histories. For example, the horned frogs (*Hemiphractus* spp.) of South America have strengthened skulls that allow them to capture and crush vertebrate prey, and the shovel-headed tree frogs (*Triprion* spp.) of Central America have hard bony projections on their head that provide protection from predators.

Most frogs have small teeth on their upper jaws, with the exception of the true toads in the family Bufonidae, who lack teeth completely. Some species have evolved teeth in their lower jaw. The aquatic Budgett's Frog (*Lepidobatrachus laevis*) of South America possesses two fanglike structures on its lower jaw, which it uses to grasp its slippery aquatic prey. Guenther's Marsupial Frog (*Gastrotheca guentheri*) from Andean Ecuador and Colombia is the only known frog to have true teeth on its lower jaw as well as its upper jaw.

The vertebral column in modern frogs has been shortened to between 5 and 8 presacral vertebrae (*Ascaphus* and *Leiopelma* have 9), compared to 24 in humans. This shortening and stiffening of the spine helps reinforce the back, enabling frogs to put more power into their locomotion. Despite frogs not having tails, they have caudal vertebrae (the vertebrae that typically make up the tail in other vertebrates) that have been fused into a rodlike structure called the urostyle. The urostyle acts as a shock absorber and is situated between the extended iliac blades of the pelvis.

Along with the shortening and strengthening of the spine, several other key adaptations have evolved to aid in jumping, the major form of locomotion used by frogs. The strengthening of bones in the limbs has enabled them to withstand the pressure from both jumping and landing. The radius and ulna are fused in the forelimb, the tibia and fibula are fused in the hindlimb, and the ankle bones (astragalus and calcaneum) are partially fused and act like a hinge joint. Several bones in the hindlimbs, feet, and ankles have also become elongated to allow for a greater jumping distance and for the attachment of powerful muscles.

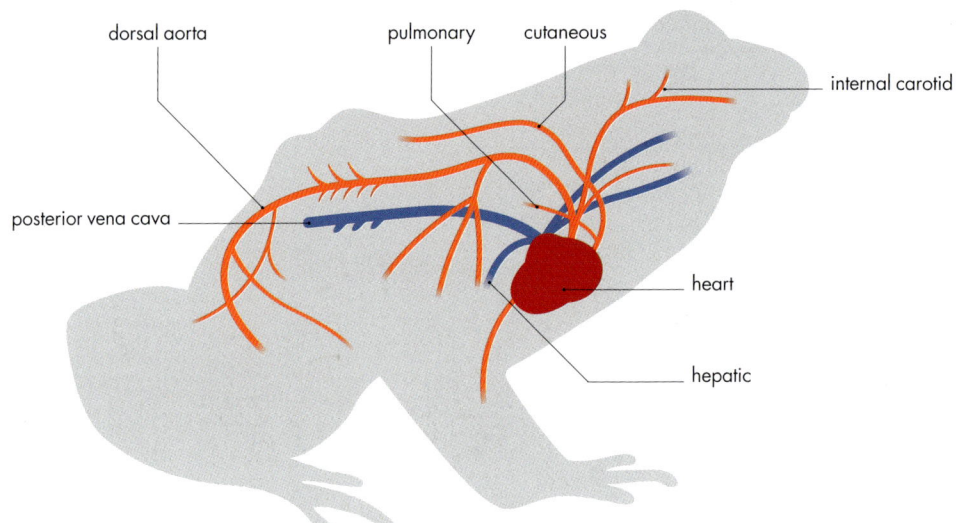

dorsal aorta

pulmonary

cutaneous

internal carotid

posterior vena cava

heart

hepatic

CIRCULATORY SYSTEM

The organs of a frog are much the same as those of other tetrapod vertebrates but with adaptations to their ectothermic lifestyle. Unlike mammals, frogs have a three-chambered heart consisting of a single ventricle and two atria.

Despite not having a ventricular septum to separate oxygen-poor and oxygen-rich blood as in mammals, birds, and crocodilians, anurans still separate the two by maintaining different pressures in the heart. These pressures direct the blood to the relevant arteries: oxygen-poor blood to the pulmocutaneous artery, and oxygen-rich blood to the carotid and systemic arteries, where it moves around the body providing oxygen to important organs.

GAS EXCHANGE

Like most other tetrapod vertebrates, frogs have paired, symmetrical lungs that are important for gas exchange within the blood. Amphibians have specialized skin that allows for efficient gas exchange with an external medium (air or water), and are one of the few vertebrates that carry

ABOVE | Major circulatory system of a frog. Red lines indicate pathways for arteries which take blood away from the heart and blue lines indicate pathways for veins returning blood to the heart.

BELOW | Cross section of a frog heart, indicating the two pathways taken by oxygen-rich blood (red) and oxygen-poor blood (blue).

HEART

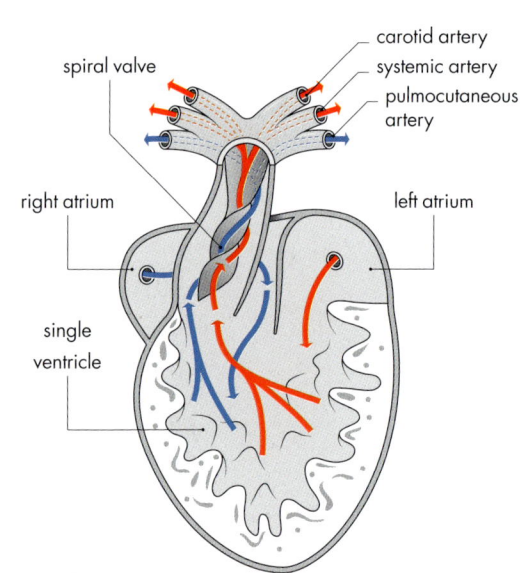

spiral valve

carotid artery

systemic artery

pulmocutaneous artery

right atrium

left atrium

single ventricle

RESPIRATORY SYSTEM

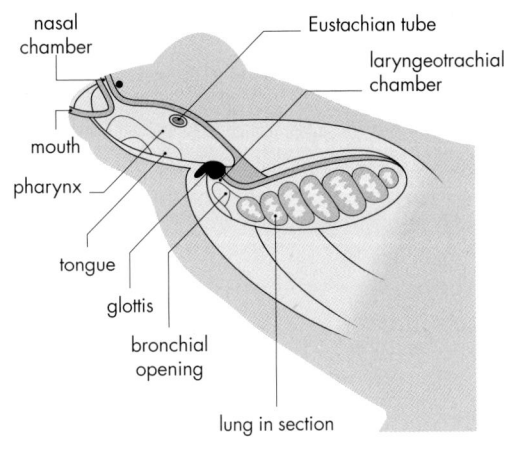

- nasal chamber
- Eustachian tube
- laryngeotrachial chamber
- mouth
- pharynx
- tongue
- glottis
- bronchial opening
- lung in section

LUNGS

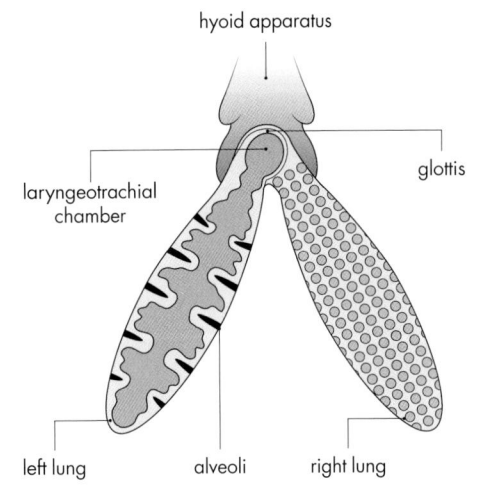

- hyoid apparatus
- laryngeotrachial chamber
- glottis
- left lung
- alveoli
- right lung

oxygen-poor blood to the skin as well as the lungs, the lungs being the normal organ for gas exchange in vertebrates. The skin of amphibians is extremely important for gas exchange and can account for 20 to 90 percent of the total oxygen intake and 30 to 90 percent of the total release of carbon dioxide. Frogs that live in an aquatic environment may use lungs predominantly for buoyancy rather than gas exchange.

Much like gills, some aquatic frogs have developed specialized ways of transferring oxygen and carbon dioxide in and out of the body. The Lake Titicaca Water Frog (*Telmatobius culeus*) from Peru and Bolivia has highly wrinkled skin, which increases surface area for gas exchange. Males of the African Hairy Frog (*Astylosternus robustus*) develop skin filaments that act like gills during the breeding season. These filaments are rich in blood vessels and increase the surface area to enable more efficient gas exchange.

TOP | Major respiratory system of a frog.

LEFT | Paired lungs of a frog. Simplified internal anatomy of the left lung.

BELOW | Cross section of frog skin. The epidermis is the outermost layer of skin that is in contact with the environment. The dermis has a high density of blood capillaries close to the surface of the skin.

SKIN

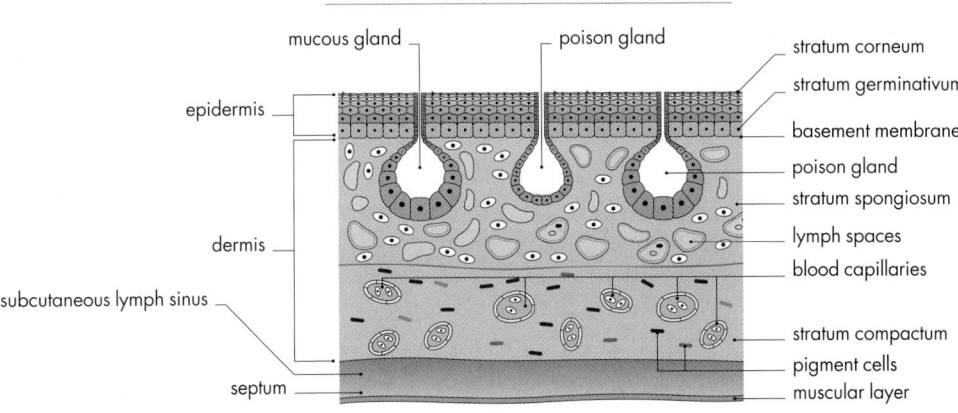

- mucous gland
- poison gland
- stratum corneum
- stratum germinativum
- epidermis
- basement membrane
- poison gland
- stratum spongiosum
- dermis
- lymph spaces
- blood capillaries
- subcutaneous lymph sinus
- stratum compactum
- pigment cells
- septum
- muscular layer

SKIN

Frog skin is an amazing structure and is used for reproduction, courtship, protection, defense, and importantly, respiration. Anuran skin is permeable and glandular and consists of an outer layer known as the epidermis and lower layer known as the dermis.

The mucus glands secrete mucoproteins, which provide the moist covering for frog skin. These mucoproteins are critical for cutaneous gas exchange through the skin. In arid conditions, several species of frog have been found to form cocoons made up of excess layers of cells that they use to estivate (dormancy during hot and/or dry periods). For example, the Northern Burrowing Frog (*Neobatrachus aquilonius*) from Australia, the Llanos Frog (*Lepidobatrachus llanensis*) from South America, and the Lowland Burrowing Treefrog (*Smilisca fodiens*) from Central America and southern Arizona have all evolved cocoon-forming capabilities.

Frogs can be brilliantly vibrant and intricate in their color patterns, even changing color throughout the day (for example, between light and dark photoperiods) and environmental conditions (for example, stress response). Chromatophores are formed of melanophores, iridophores, and xanthophores. The ways that the different chromatophores interact with each other account for the diversity of colors and color changes that we see in frogs.

Water loss is a major concern for frogs due to skin permeability. Water balance is maintained using behavioral and physiological adaptations.

Behavioral adaptations can include moving into water or retreating into cooler and damper areas such as under fallen logs. Many species will curl their bodies into tight balls when humidity is low, thus reducing surface area and consequently evaporative water loss. In hot conditions, but where water might be readily available, some species may balance their temperature with evaporative water loss, as seen in the diurnal Rockhole Frog (*Litoria meiriana*) from the Kimberley region of Western Australia and the Northern Territory, Australia, which is active in the middle of the day when predators need to escape the heat.

The glass frogs of the family Centrolenidae have translucent belly skin, hence their name. When viewed from underneath, the organs of the frogs can be clearly observed and you can even see the heart beating! The translucent bellies are to aid with camouflage using a phenomenon known as edge diffusion, which creates a color gradient from leaf to frog making it harder for a predator to locate the frog.

OPPOSITE | The organs of the Reticulated Glass Frog (*Hyalinobatrachium valerioi*) can clearly be seen through the belly skin. It is even possible to watch the heart beating in live animals!

BELOW | The Lowland Burrowing Treefrog (*Smilisca fodiens*) folds its limbs tightly against its body during cocoon formation to reduce surface area, therefore decreasing evaporative water loss.

SENSES

VISION AND EYES

The eyes of frogs are some of the most spectacular in the world, with ripples, spots, stars, lines, and huge variations in color. Frog pupils (the dark opening in the center of an eye) are also highly variable, with some species having round holes, vertical slits, horizontal slits, diamond-shaped pupils, or star-shaped pupils. Pupils control the amount of light entering the eye and are usually linked to the ecology of the species. For example, the nocturnal Red-eyed Tree Frog (*Agalychnis callidryas*) from Central America and Colombia has vertical pupils when seen in the day (thus protecting it from excessive UV), but during dark nights the pupils expand to a near-complete circle, enabling as much light as possible to enter the eye.

Most frogs have large eyes that protrude above the top of their heads, giving them extremely wide fields of vision. The size of the eyes and several other key adaptations mean that frogs have excellent vision and accurate depth perception. However, not all frogs have large eyes, with members of the Microhylidae, Myobatrachidae, Rhinophrynidae, and Pipidae having reduced eyes, which in most species are associated with a burrowing lifestyle, living in leaf-litter or a murky aquatic environment. Having large eyes in such an environment has no advantage and may even be a burden.

Unlike humans, frogs have three distinct eyelids. They have the usual upper and lower eyelids and a third one known as a nictitating membrane. Nictitating membranes (also seen in

birds and crocodiles) are transparent and help protect the eye from damage and drying out. In an aquatic environment they act like goggles, allowing the frogs to see underwater without water coming into direct contact with the eyeball.

The retina of vertebrates contains two types of photoreceptor cells—rods and cones—that sit at the back of the eye and transmit signals to the brain where an image is formed. Rods are generally used in low light (scotopic) levels and cones are used in bright light (photopic) conditions. In vertebrates, cones are associated with color vision, with different types associated with specific wavelengths. Nocturnal species tend to have more rods and diurnal species have more cones. Due to the damage that UV light can cause to the retina, diurnal frogs tend to have UV-blocking pigments in their lenses. The diurnally active Boraceia Tree Toad (*Hylodes phyllodes*) from Brazil relies on visual stimuli between conspecifics (members of the same species), as it lives near to fast-flowing streams and thus excellent vision is critical. A UV-protective lens means that an individual's vision will not deteriorate through time due to UV radiation.

Frog eyes are extremely sensitive, with many species being able to navigate perfectly through their environment in extremely low light levels. This is an important adaptation for many species that are nocturnal and is especially true for those that are nocturnal and live in forests or leaf-litter, where it is even darker. Unlike most other vertebrates, frogs have two types of rods, which enables them to see color even in very low light levels. The "green rods" of frogs (and some salamanders) absorb lower wavelengths than that of typical "red rods," enabling them to have nocturnal color vision.

ABOVE | Examples of some of the amazing diversity of frog eyes.

OPPOSITE | From southwestern Colombia and northwestern Ecuador, the Imbabura Treefrog (*Boana picturata*) has impressively large eyes.

VOCALIZATION

Male frogs are fantastic vocalists and use calls to attract females, assert dominance, and establish territories. Both males and females can also produce sounds to warn off predators or as release calls if an overzealous male tries to mate with another male or an already mated female. Sounds produced by frogs can be simple pips to complex calls consisting of several different types of frequencies, rates, and timing. The majority of frog calls that you hear are male frogs singing a love song to attract a female.

Most male frogs have a single or paired vocal sacs that amplify their calls to sometimes staggering levels. The tiny Puerto Rican Coqui (*Eleutherodactylus coqui*) has been reported to call at 95 dB, the same volume as a loud motorcycle engine. For a small frog of just 1¼ in (34 mm), that is an incredible amount of noise! On the other hand, pipid frogs have no vocal sacs or vocal cords but make clicking sounds by pulling two sides of the larynx apart, which contracts the laryngeal muscles.

Male frogs typically use external vocal sacs to amplify the volume of their calls when inflated. Most frogs have a single vocal sac that can be seen underneath the chin, in an area known as the buccal cavity. Some frogs, however, have two vocal sacs, with two different types occuring. Paired throat sacs sit in the same area as the single throat sac but when inflated stick out slightly wider than the width of the frog's head. Lateral sacs sit at the

rear of the jaw and when inflated seem to project almost perpendicular to the head. All vocal sac types have paired slits opening into the buccal cavity.

The Central Coast Stubfoot Toad (*Atelopus franciscus*) from French Guiana lives in very loud microhabitats at the edge of fast-flowing water. The species has no tympanum (external eardrum) and no external vocal sac. Despite this, males still call but the sound only reaches less than 26 ft (8 m) and is likely used only for securing their territory. The Hole-in-the-head Frog (*Huia cavitympanum*) from Borneo and Concave-eared Torrent Frog (*Odorrana tormota*) from China get around the problem of living near fast-flowing water by having high-frequency and ultrasonic calls that cannot be detected by humans.

HEARING

Frogs have very good hearing, which has evolved primarily because their main form of communication is auditory. Frogs have an external eardrum known as a tympanum, which appears as an oval disc of slightly different-colored skin situated just behind the eye on each side of the head. The tympanum is a tightly stretched membrane, connected by a ring of cartilage, and acts like the skin of a drum. It receives sounds in the forms of vibrations through the air and transfers this to the inner ear, where small hairlike structures called "hair cells" start to move, triggering nerve fibers to carry electrical impulses to the brain, where the sounds are interpreted.

ABOVE | Green Frogs (*Aquarana clamitans*) from North America have very large tympanums. The tympanum is the large disc located directly behind the eye.

OPPOSITE | "Earless" Oriental Fire-bellied Toads (*Bombina orientalis*) use their lungs to pick up mostly low-frequency sounds in the absence of a tympanum.

Amazingly, frogs have one unique adaptation: they use their lungs to help assist with the processing of sounds. Frog lungs have a direct link with the inner ear via the eustachian tubes. This adaptation likely came about to avoid injuring their own eardrums with their loud calls. They achieve this by equalizing the pressure between the outer surface of the eardrum and the inner surface.

Some frogs are known as "earless frogs" because they lack an external tympanum, including members of the Central and South American *Atelopus* and the Oriental Fire-bellied Toad (*Bombina orientalis*) from northern Asia. The inner ears of *Atelopus* have the normal inner ear configuration, and amazingly they can pick up sounds at the same level as frogs that have a tympanum.

BELOW | For frogs, the most important sense is hearing, and many species possess extremely large external eardrums known as tympana (singular: tympanum) which may be larger than their eyes, while other species lack external tympana, and still others exhibit internal tympana connected to the outside world by auditory canals similar to our own.

EAR

inner ear | middle ear

otic capsule | tympanic membrane

saccule

periotic sac

stapes

basilar papilla | operculum

amphibian papilla

REPRODUCTION

When it comes to reproductive strategies, frogs have exceptional levels of diversity, with 60 or more different modes reported. Mate selection is usually done by the female, who will choose a male based on different criteria depending on the species, such as breeding site or the sounds or physical attributes of the male. Almost all frogs use external fertilization, whereby the female will release unfertilized eggs from her cloaca and the male will cover the eggs in sperm in order to fertilize them.

BELOW | Schematic representation of female (left) and male (right) frog reproductive systems.

REPRODUCTIVE SYSTEMS

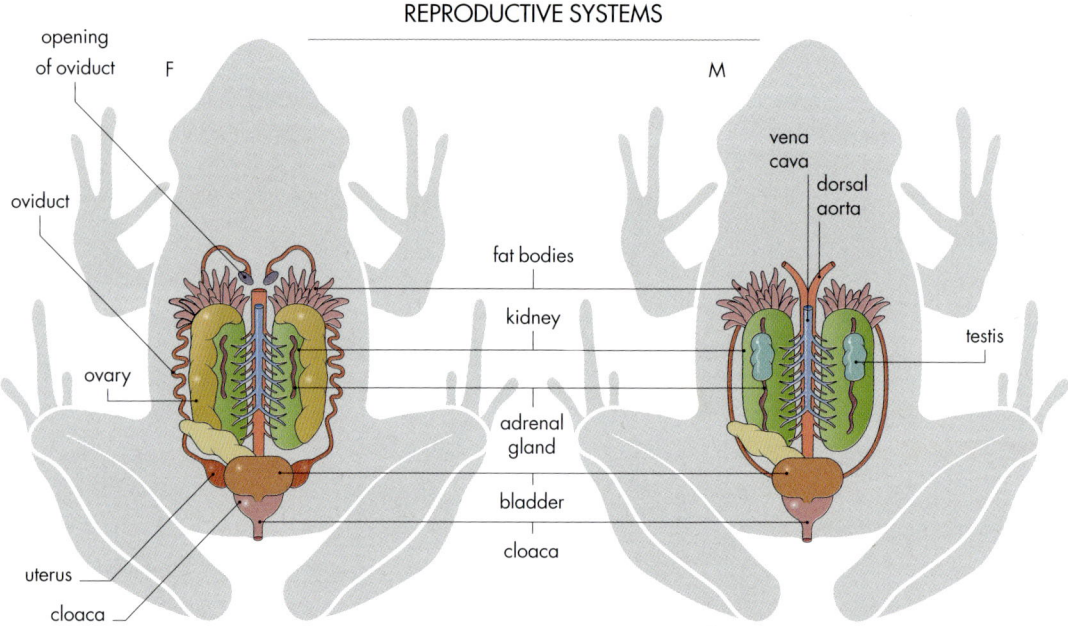

F
M

opening of oviduct
oviduct
ovary
uterus
cloaca

vena cava
dorsal aorta
fat bodies
kidney
testis
adrenal gland
bladder
cloaca

SEXUAL DIMORPHISM AND SEXUAL DICHROMATISM

How do we tell male and female frogs apart? This is not an easy question to answer, because within the more than 7,600 species of frogs, there are lots of exceptions to the rule. Typically, the female tends to be larger than the male (likely because she will need to accommodate the eggs during development). The male, on the other hand, develops nuptial pads on his forehands (specialized skin glands that swell when triggered by breeding hormones), sometimes has thicker forearms, and is often the only sex to vocalize when trying to attract a mate. Males vocalize by using a vocal sac and this is often still visible when not inflated as wrinkly or baggy skin, usually under the chin.

Many frogs have profound differences in color between the sexes; this is known as sexual

ABOVE | Male Phinda Rain Frogs
(*Breviceps carruthersi*) are significantly
smaller than females. Due to the size
difference, the species uses sticky
secretions from the skin to "glue"
themselves together.

INSET | Males of the Stony Creek Frog
(*Ranoidea wilcoxii*) are much smaller than
the females and breeding males can
rapidly change color to bright yellow
during amplexus.

dichromatism. Some species only become sexually
dichromatic for a short period during the breeding season.
Males of the Yellow Toad (*Incilius luetkenii*) undergo a color
change for just a few hours directly prior to breeding; this
behavior is known as dynamic dichromatism. Other species
have ontogenetic dichromatism, whereby the differences are
present permanently following metamorphosis, such as the
Argus Reed Frog (*Hyperolius argus*), which varies in both color
and pattern between the sexes.

AMPLEXUS POSITIONS

INGUINAL

AXILLARY

INDEPENDENT

DORSAL STRADDLE

CEPHALIC

HEAD STRADDLE

GLUED

EXTERNAL FERTILIZATION

To mate, most frogs perform a behavior known as amplexus. This typically involves the male grasping the female in order to navigate his cloaca over hers so that when she releases her eggs, he can release his sperm. During the breeding season, males will often develop nuptial pads at the base of their thumbs, which help to clasp the females during amplexus.

There are huge variations in the mating position of frogs, with no less than seven different amplexus positions, referred to as axillary, inguinal, cephalic, head straddle, glued, dorsal straddle, and independent. Amplexus can be exceptionally brief (just a few seconds) or up to several months, depending on the species.

Axillary and inguinal are the most widespread forms of amplexus. During axillary amplexus, males grasp the female below her armpits, whereas in inguinal amplexus the male will grasp the female around her waist. Cephalic amplexus, similar to axillary and inguinal amplexus, is where males will instead grab onto the female's head.

Other methods exist whereby the male will not physically grasp the female. Head straddle amplexus involves the male straddling the female's head from behind, and when she releases her eggs he releases his sperm, which trickles down her back and onto the unfertilized eggs. During dorsal straddle amplexus the male pushes the female onto a substrate but without clasping her body.

Some species that have incredible sexual size dimorphism, such as rain frogs (*Breviceps* spp.), have had to adapt to a somewhat impossible challenge. Males are dwarfed by the females and only have tiny arms, making other forms of amplexus impossible. These frogs have evolved an intriguing mechanism in order to mate, called glued amplexus. Males produce a "glue" on their ventral surface and females produce an "adhesive" on their backs. Males will climb onto the backs of the females and literally stick themselves to the females using this glue.

INTERNAL FERTILIZATION

The two species of North American tailed frogs (Ascaphidae) are the only frogs to have a specialized appendage for internal fertilization. The male has a modified cloaca that forms what is known as an intromittent organ, which he uses for internal fertilization by inserting it into the female's cloaca during mating. The female can store sperm internally for up to nine months, until conditions are suitable for her to lay eggs.

There are a handful of other frogs that have internal fertilization, but this is not achieved with an intromittent organ like with the tailed frogs.

Instead, the male and female will face in opposite directions (independent amplexus) and usually press their cloacas together (cloacal apposition). The male will then release sperm into the female and fertilization follows.

Most species with internal fertilization are viviparous, meaning that they do not lay eggs. The Sulawesi Fanged Frog (*Limnonectes larvaepartus*) is one of these, giving birth to tadpoles, and is the only species known to do this. Some African toads in the genera *Nectophrynoides* and *Nimbaphrynoides* (known as viviparous toads) and the Golden Coqui (*Eleutherodactylus jasperi*) from Puerto Rico retain eggs internally before giving birth to froglets, which are perfect mini replicas of the adults. They provide nourishment to the developing young via epithelial (cells that line the inner and outer regions of the body) secretions within the oviduct.

OPPOSITE | Diagram representing the seven major mating positions of frogs.

RIGHT | The intromittent organ of the Coastal Tailed Frog (*Ascaphus truei*) is clearly visible between the two hindlimbs.

COMPETITION

There is often fierce competition when trying to secure a mate, with males usually having to compete in a number of different ways to gain the favor of a female. Large mating aggregations are common, especially in still or slow-flowing water breeders, where male–male competition can be high. Some frog species will defend breeding sites ferociously, fighting with rivals. The world's largest frog, the Goliath Frog (*Conraua goliath*), will fight so savagely that it often leads to the deaths of other hopeful breeding males.

Most species are not so violent and use more passive ways of attracting a female, calling being the most obvious method. As well as being a warning about their toxicity, the vivid coloration of poison dart frogs (Dendrobatidae) may also play a part in mate selection. Some frogs have adopted unusual visual signaling to assert dominance, territoriality, and/or attract a female. Most of these frogs live near fast-flowing streams, rivers, or waterfalls, where vocalizations might be obscured by the noise of running water. One example of this is Orejuela's Glassfrog (*Sachatamia orejuela*) from Ecuador and Colombia, where males flap their feet, wave their hands, and bob their heads to attract females. This behavior is also seen in the Indian dancing frogs (*Micrixalus*).

Once a male has attracted a female, he will often defend her from rival males by remaining in close proximity, often staying in amplexus for prolonged periods. Males of the Santa Marta Harlequin Toad (*Atelopus laetissimus*) from Colombia will stay in amplexus with the female for a month or more and competition can be fierce, with other rival males attempting to displace the amplectant male. Amplectant male displacement is very common in some species; for example, approximately one-third of displacement attempts in the European Common Toad (*Bufo bufo*) are successful.

This fierce competition can often lead to mating balls, where large numbers of males attempt to displace other males. Sometimes these mating balls can involve tens of males

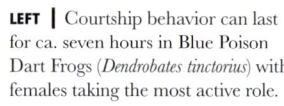

LEFT | Courtship behavior can last for ca. seven hours in Blue Poison Dart Frogs (*Dendrobates tinctorius*) with females taking the most active role.

OPPOSITE | Common Toads (*Bufo bufo*) form large mating balls during their explosive breeding season.

with a single female. Unfortunately, these mating balls can have deadly consequences, with both males and females often drowning in the process.

Other species have sneaky males in their midst. For example, European Common Frog (*Rana temporaria*) males that have not been successful in mating will sometimes crawl into egg masses and release their sperm. Due to the volume of eggs produced by a single female, the amplectant males cannot fertilize all of the eggs (especially those in the center of the egg mass). These sneaky males will therefore get to fertilize the eggs, which is beneficial to both the female and the non-amplectant male.

EGG-LAYING

Frogs lay eggs in a number of different ways. Some species lay truly astounding numbers of eggs, such as the Cane Toad (*Rhinella marina*), where a single female can lay upward of 35,000 eggs. Frogs that lay eggs in water typically have two main forms, large egg masses (clumps) and strings (mostly produced by toads).

Some frog species, especially in the tropics, do not lay their eggs in the water. Many larvae of treefrogs still require an aquatic environment once they hatch, and so they are laid on leaves or rocks directly above the water. The Seychelles Treefrog (*Tachycnemis seychellensis*) is a highly adaptable species, laying its eggs on overhanging leaves, rocks, or fully submerged in the water.

In regions without large bodies of water, such as in rainforests, frogs must adapt to these conditions. Despite there being a seeming lack of still water sources in those environments, there are plants that fill this niche. Bromeliads have permanent water in their crown (center), known as phytotelmata or tanks. These phytotelmata are used by an array of animals, including frogs. The Bahia's Broad-snout

Casque-headed Tree Frog (*Nyctimantis arapapa*) from the Brazilian Atlantic Forest lays eggs in bromeliads, with males attracting females by calling from inside the tank. Once the eggs have been fertilized, the male will protect them (and subsequent tadpoles) by forming a seal just above the tank using his highly ossified skull. This behavior is known as phragmosis.

GOOD PARENTS

Following egg-laying, some frogs simply leave their offspring to their own fate, whereas others are exceptional parents. Many species are known to carry eggs on their backs, which provides the vulnerable eggs with added protection. After the fertilization of eggs, male midwife toads (*Alytes* spp.) from Europe and northwest Africa attach the eggs to their lower backs and carry them until the tadpoles hatch.

Females of the Hemiphractidae family also look after their eggs (and tadpoles or froglets) on their backs. Many of the marsupial frogs (*Gastropheca* and *Flectonotus* spp.), egg-brooding frogs (*Fritziana* spp.), and horned treefrogs (*Hemiphractus* spp.) have perhaps the most developed means of doing this, having a specialized dorsal brood pouch known as a marsupium. Unlike most frogs that have lecithotrophic (no nutrition received other than from the yolk) eggs, the embryos inside brood pouches may receive nutrients during development due to the highly vascularized nature of the pouch. Males of the Australian myobatrachid Pouched Frog (*Assa darlingtoni*) look after their tadpoles in "pockets" on each side of the body before they emerge as froglets.

The Suriname Toad (*Pipa pipa*) has an incredible adaptation whereby the female

will release one egg at a time, which the male fertilizes and attaches to her back. The female's skin starts to become thicker and thicker until it envelops the 100 or so eggs. All of the developmental stages of the young happen within the skin, until eventually the froglets erupt out of the female's back! Once all of the young have emerged, the female will shed the thicker layer of skin that she developed.

The now-extinct gastric brooding frogs (*Rheobatrachus* spp.) from Australia had another unique way of protecting their eggs. The female would swallow the fertilized eggs, which then sat in the stomach until fully developed froglets were regurgitated. To do this, the female's digestive system would shut down. Her stomach would become so engorged that the lungs could no longer function, and so full gas transfer had to occur directly through her skin!

FEEDING AND DIET

OPPOSITE | American Green Tree Frog (*Dryophytes cinereus*) capturing a Hawk Moth using its tongue.

Most adult frogs are carnivorous and tend to eat small invertebrate prey, whereas tadpoles are mostly herbivorous or nonfeeding. As with almost all aspects of frog life histories, there are exceptions to the rules and some species consume relatively large vertebrate prey.

AQUATIC FEEDING

Most tadpoles are filter-feeders. They suck in water through their mouths, where particles are captured by sticky mucus or food traps and consumed. These particles consist of plant material that tadpoles scrape off surfaces using their keratinized mouth parts, or free-floating particles such as phytoplankton. The tadpoles lift the ceratohyal cartilage, which causes the orbitohyal muscle to drop, increasing the buccal volume substantially, which draws in water from outside. This water then passes over the gills and out of the spiracle, and thus this behavior aids in both feeding and gas exchange.

Tadpoles' mouthparts match their feeding ecologies. Generalist feeders have keratinized beaks that are used for scraping. Surface feeders have mouths that are upward-pointing so that they can feed on particles on the water's surface, such as the Asian Horned Frog (*Megophrys montana*). Suspension feeders stay in the middle of the water column and have a terminal (pointing forward) mouth, such as the African Clawed Frog (*Xenopus laevis*). Some species have mouth parts that allow them to stick to and feed on substrates in fast-flowing streams without being washed away, or have specialized suction cups on their bellies, such as the Mountain Cascade Frog (*Sumaterana montana*) from Sumatra. Carnivorous tadpoles often have large denticles (toothlike projections) that they use to rip flesh off other tadpoles or soft-bodied prey.

Aquatic adult frogs will often use suction-feeding like tadpoles but to capture prey instead

of particulates in the water. Like other aquatic vertebrates, they use inertial suction, which sucks prey into their mouths. This behavior is common in members of the Pipidae from Africa and South America. Some pipid species will use their forearms to help push large prey items into their mouths.

TERRESTRIAL FEEDING

Most adult frogs use tongue projection to capture prey. The mechanisms that are used to do this vary species by species. Some frogs that split from other species early in the evolution of frogs have slow projection speeds and the tongues only extend just beyond the front of the mouth, as in the midwife toads (Alytidae) of Europe and North Africa and the tailed frogs (Ascaphidae) of North America. Like the pipids, they will often use their forearms to help push prey into their mouths.

Other frogs use a catapult-type structure in the mouth, where the tongue is flung from the mouth in a process known as inertial elongation. The tongue acts as an elastic band and is tensioned before being released. While this motion is fast,

it lacks some accuracy due to the level of extension that is often seen (up to nearly twice the length of the frog's head). This mechanism is used to catch small invertebrate prey.

The slower hydrostatic tongue protrusion is used by ant and termite specialists. While the extension of the tongue is slower, it is incredibly accurate. Other species are voracious and will eat other vertebrates (sometimes the same size as themselves), such as the South American horned frogs (*Ceratophrys*), which are capable of holding onto prey employing a pressure up to six and a half times their own weight!

Once frogs have captured and subdued their prey, they use a fascinating mechanism to swallow. Have you ever noticed that when a frog closes its eyes, the once-protruding eyes sink into the head? If you think back to the skull structure, you will note that the orbit (eye) holes are large and have no bone structure below them. Frogs take advantage of this anatomical feature. The eyes are pulled down into the mouth, which pushes the food down the esophagus.

ABOVE | Amazonian Horned Frogs (*Ceratophrys cornuta*) are voracious predators and will even feed on other frogs.

DEFENSE

At first, frogs may seem like defenseless animals with few adaptations to help protect them from predators and microorganisms, reptiles, other amphibians, mammals, fish, birds, and invertebrates. However, frogs have evolved some spectacular physiological and behavioral adaptations that have helped them survive for hundreds of millions of years against predators, besides simply hopping away or puffing up to make themselves look larger.

Some frogs will do what we would consider to be a typical vertebrate response to being threatened: they will bite. Biting in many species would not be effective due to small teeth or weak jaws, but some species have powerful jaws and do not hesitate to use them. Horned ("Pac-Man") Frogs (*Ceratophrys* spp.), for example, have a bite force that is able to restrain and immobilize prey the same size as themselves.

As we have seen earlier in this chapter, frogs are amazing vocalists. They use calls to not only signal to mates or to warn conspecifics that it is their territory, but also to deter predators. If you have ever picked up a frog, it might have made a noise, which is a warning for you to release it. Rain Frogs (*Breviceps* spp.) take this one step farther and when threatened make a comical, surprisingly loud screeching sound, while puffing up their body to make themselves look larger.

POISONOUS FROGS

Almost all amphibians produce toxins of some form and can either sequester these from their diet, produce them directly through their skin glands (or other specialized glands), or gain them from relationships with symbiotic microorganisms. These compounds are extremely diverse, with more than 900 different alkaloids that have been identified in frogs. Most frogs produce antimicrobial peptides to protect them from harmful microbial communities. Some species of frog produce exceptionally strong toxins, capable of killing a human tens of times over. The most toxic frog is the Golden Dart-poison Frog (*Phyllobates terribilis*) from Colombia. This species produces the steroidal alkaloids Batrachotoxin, Batrachotoxin A, and

Homobatrachotoxin. While most animal toxins that are not injected (and thus not classified as venoms) do not have any effects on the human body unless they are ingested or enter the bloodstream, the toxins of the Golden Dart-poison Frog are so powerful that they can cause a strong burning sensation for hours if they make contact with the skin.

Toads (Bufonidae, Pelobatidae), forest frogs (Odontophrynidae), and treefrogs (Hylidae) have evolved specialized glands on either side of their neck, called parotoid (or parotid) glands. When harassed, true toads will produce a thick, viscous white toxin, which is distasteful or even deadly to organisms that eat them. One of the most invasive amphibians, the Cane Toad (*Rhinella marina*), has had catastrophic impacts in many regions where it has been introduced. This is most prominent in Australia, where it has decimated many reptile species that preyed on native amphibians. There are no native toads in Australia and so many of the naïve native species have not evolved to be able to cope with their toxins. Some species, such as the ranid Malabar Frog (*Clinotarsus curtipes*) from the Western Ghats of India, have tadpoles with parotoid glands.

Members of the poison dart frogs (Dendrobatidae), golden frogs (Mantellidae),

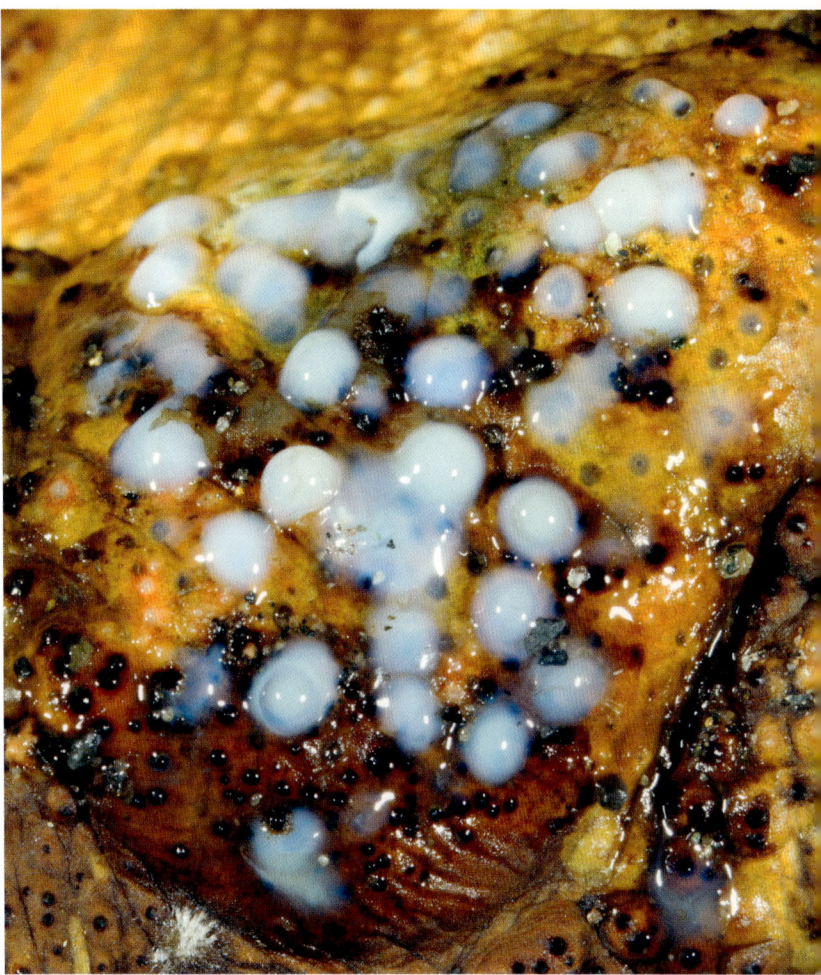

Australian froglets (Myobatrachidae), South American redbelly toads (Melanophryniscus), and some robber frogs (Eleutherodactylus) acquire some highly toxic alkaloids from their arthropod (invertebrate) prey rather than producing it themselves. Because they rely on specific prey types for obtaining most of their poisons, in captivity, where their diet mostly consists of farmed crickets and flies, they quickly lose their toxicity.

VENOMOUS FROGS

Poisons and venoms are not interchangeable terms, because by definition venom is injected into the body whereas poisons are consumed, absorbed, or inhaled. Bruno's Casque-headed Frog (*Nyctimantis brunoi*) and Greening's Frog (*Corythomantis greeningi*) have evolved an amazing adaptation to turn their poisonous skin secretions into venom. When attacked, both species will headbutt their attacker, which causes bony projections on their skull to pierce through their own skin. These bony projections are then covered in their toxic skin secretions before they pierce their attacker, thus injecting venom through the wounds.

CAMOUFLAGE AND CRYPSIS

Some of the camouflage capabilities of frogs are truly astounding, with some species blending in perfectly with their background. Perhaps one of the best examples of this can be found in the Vietnamese Mossy Frog (*Theloderma corticale*), which looks remarkably like moss. Not only is their color speckled in a dark to light green and brown, but they have tuberculate skin (having small, rounded projections) and their eye pattern and color even matches their skin. Other species have evolved to match even unexpected features in their environment, such as the South American hylid *Dendropsophus marmoratus*, which resembles bird poo!

Aquatic frogs have also had to adapt to blend into their environment. The Suriname Toad (*Pipa pipa*) has evolved to look very much like a dead leaf sitting at the bottom of slow-flowing rivers. They have triangular-shaped heads, a flattened body, are brown/olive, and have tubercles along their skin. They will rest with their arms to the sides of their body and remain motionless for extended periods of time.

Some species opt to confuse predators and use a disruptive coloration tactic to do this. They achieve this by using colors and patterns to break up their body outline. The most common ways are by incorporation of blotches or a dorsal line.

ABOVE LEFT | Suriname Toads (*Pipa pipa*) will rest motionless in slow moving water waiting for prey to swim close enough for them to capture. Their body form makes it difficult for predators to spot them.

ABOVE RIGHT | Vietnamese Mossy Frogs (*Theloderma corticale*) display incredible camouflage when resting on mossy substrates.

COLOR AS A WARNING

Many species of frog are vibrant in color. This may seem counterintuitive from a predation perspective because they will be highly visible. However, these bright colors act as warning signals to would-be predators about the toxic nature of the frogs; this toxic advertisement is known as aposematism. No frogs exemplify this more than the Central and South American poison dart frogs (Dendrobatidae), cryptic dart frogs (Aromobatidae), and harlequin toads (*Atelopus* spp.); the Atlantic Forest of South America's saddleback toads (*Brachycephalus* spp.); and Madagascar's tomato frogs (*Dyscophus* spp.) and mantellas (Mantellidae).

Other species do not continuously display their aposematic coloration and instead conceal it when resting. Fire-bellied toads (*Bombina* spp.) only have aposematic coloration on their ventral surface and when threatened they will lift their body off the ground, displaying their bright-red underside, a tactic known as unkenreflex.

Other species have eye spots, such as the Brazilian leptodactylid frog *Physalaemus deimaticus*, which when threatened raises up its rear end and puffs up its body, displaying dramatic black spots that are thought to mimic the eyes of snakes. This behavior is known as a "deimatic" display.

TOP | The Yellow-banded Poison Frog (*Dendrobates leucomelas*) has bright coloration to warn predators of its toxicity.

MIDDLE | The red coloration of Tomato Frogs (*Dyscophus antongilii*) is a warning that when harassed they will secrete a sticky toxic substance from their skin.

RIGHT | The exceptionally brightly colored Purple Harlequin Toad (*Atelopus barbotini*) uses its colors as a signal about its toxicity.

OTHER ADAPTATIONS

There are lots of other mechanisms that frogs use to avoid predation. When attacked, the Hairy Frog and night frogs (*Astylosternus* spp.) from Central Africa will break their bones in their hind feet and then push these through their own skin. They then use these protruding bones to stab into the attacking animal. This adaptation has given *Astylosternus robustus* the alternate name of the Wolverine Frog, in reference to the claws that emerge from the hands of the Marvel superhero Wolverine.

Many frogs will urinate when harassed. This urine can be smelly and foul-tasting, meaning that if a predator has the frog in its mouth, it might be tempted to release it. Other species may bury themselves in mud or feign death by remaining motionless to avoid confrontations with predators in the first place. The Pebble Toad (*Oreophrynella nigra*) from northern South America will curl up into a ball and roll away from danger.

USING OTHER ANIMALS FOR PROTECTION

Some frogs use other animals as protection. For example, members of the species-rich narrow-mouthed frogs (Microhylidae) from South America and Asia live with spiders. These small frogs gain protection from these often large spiders, and in return it is believed that the frogs eat the small arthropods that would otherwise cause problems

ABOVE | A Dotted Humming Frog (*Chiasmocleis ventrimaculata*) coexisting in a mutualistic relationship with a large Peruvian Tarauntula (*Pamphobeteus* sp.).

LEFT | The bones on the hindfoot of the Wolverine Frog (*Astylosternus robustus*) have been pushed through the skin as a defense mechanism.

for the spiders. This is a mutualistic behavior because both the frog and spider benefit from the relationship. It is possible that the relationships between the spiders and frogs are species-specific. The Dotted Humming Frog (*Chiasmocleis ventrimaculata*) from the western Amazon region of South America is known to co-occur with several species of tarantula and other spiders. The spiders use chemical signals to recognize these frogs instead of eating them, as they have been recorded to do for other species of frog.

LOCOMOTION

Frogs move in a number of different ways but the most well known is hopping or jumping. Almost everywhere that frogs occur there will be at least one species that hops or jumps. Frogs are particularly adept at jumping due to their enlarged hind limbs and strong thigh muscles. These large limbs are capable of propelling the frogs forward, often over incredibly large distances for their size. The American Bullfrog (*Aquarana catesbeiana*) is a large frog at about 7 in (180 mm) in length, but despite this the species has a truly impressive jumping capability, recorded at up to 7 ft 2 in (2.2 m), which is a staggering 12 times its total length. However, the American Bullfrog's leap looks miniscule considering its body size when comparing it to the Sharp-nosed Grass Frog (*Ptychadena oxyrhynchus*) from southern Africa, where one

individual jumped 17 ft 6 in (5.35 m), which was more than 90 times its total length! Although not all species are capable of such large leaps, other species are capable of extreme speed. The Coastal Ecuador Smoky Jungle Frog (*Leptodactylus peritoaktites*) when disturbed is incredibly fast, combining powerful leaps one after the other. The speed is so great that even on flat ground it is often difficult for a human to keep up when running from a standing start.

WALKING AND RUNNING

Not all frogs jump when navigating through their environment, with some opting to walk or run instead. All frogs are capable of walking, but it is often cumbersome for many species due to their large hindlimbs. However, some species are better adapted to walking and running than others, with

these species often having shorter hindlimbs than those of other frogs, meaning that their hindlimbs are more similar in length to their forelimbs. One species that almost exclusively moves in this way is the European Natterjack Toad (*Epidalea calamita*), which is capable of running at fairly quick speeds. This, coupled with the "go-faster stripe" that tracks down the center of its dorsum, gives the species its other common name of Running Toad.

SWIMMING

As discussed earlier, tadpoles are well adapted to an aquatic lifestyle, but many adult frog species are also adapted to life in an aquatic environment. Species that have some association with water courses generally have webbed feet to enable them to swim more efficiently. The level of webbing usually depends upon the extent to which the species is linked with water—for example, species that are fully aquatic have extensively webbed feet. The African Clawed Frog (*Xenopus laevis,*) has large webbing on its hindfeet that it fans out and uses to make big paddle-like strokes while moving. Perhaps the most interesting aquatic adaptation of any frog can be seen in the Hairy Frog (*Astylosternus robustus*) from West Africa, where there is extreme sexual dimorphism and the males spend most of their lives in the water. The males have papillae on the sides and thighs, increasing their surface area for highly efficient gas exchange while in the water.

ABOVE | A European Tree Frog (*Hyla arborea*) displaying a typical jumping motion of a frog.

CLIMBING

Frogs have adapted equally well to life in trees as they have to living in water and on the ground. Arboreality has evolved numerous times, to occupy the countless niches that life in the trees offers. Throughout the tropics, tree-dwelling frogs are as abundant as frogs that live on the ground. Arboreal frogs have had to adapt to ensure that they are capable of supporting their body weight in the trees, regularly achieving this by having adhesive toe pads that are often wider than their ground-dwelling counterparts, a lengthening of their fore- and hindlimbs, and increased flexibility in their ankle joints.

GLIDING

There are several species of frog that can sail through the air to escape predators or to move through their environment with more energy efficiency, in a process termed parachuting or gliding. This behavior is especially prevalent in members of the Flying Frogs (*Rhacophorus*). These frogs have extensive webbing between their digits that act like

parachutes when the frogs jump from a high point. The skin flaps, patagia, are extended during flight, increasing the surface area and enabling the frog to slow its descent and glide through the air.

BURROWING

Burrowing is a common feature among frogs, scattered across several families. Almost all burrowing species will use the metatarsal tubercles on their hind feet, which act like shovels, to excavate the soil. Other species will burrow forward. For example, the Shovelnose Frogs (*Hemisus* spp.) of Africa burrow headfirst by using their shovel-shaped nose to push their way through the substrate.

Different species use burrowing behavior in different ways. Some will use the behavior to escape or hide from predators, some will burrow to retain moisture, some dig burrows in order to breed or rear young, and others will forage for soil-dwelling invertebrates.

LIFE IN EXTREME CONDITIONS

Frogs have a global distribution, occurring on every continent except Antarctica. They can be found in almost every terrestrial habitat on earth, and in many temperate regions they are considered ubiquitous within wetland habitats. However, frogs have an incredible ability to adapt when we consider that their anatomy and physiology mean they are tied to water and are ectothermic. As discussed on page 18, some species can form cocoons to estivate during extreme conditions.

Frogs do not cope well with salt water due to their need to use their skin for gas exchange, and salt prohibits this. This is one reason why amphibians are rare on geologically recent islands. Some families seem to tolerate transoceanic dispersal, such as the Hyperoliidae of Africa, which have colonized islands to the west and east of the main landmass, including as far as the Seychelles archipelago. The Cane Toad (*Rhinella marina*) can tolerate saline conditions and has even been observed swimming in the sea on occasion. The Crab-eating Frog (*Fejervarya cancrivora*) of Southeast Asia lives in brackish environments and eats a range of crustaceans (for example, crabs) and insects that inhabit these environments.

Many frogs hibernate to escape the cold temperatures of winter. Perhaps the most amazing frog species to have adapted to extreme cold conditions is the Wood Frog (*Boreorana sylvatica*) of North America, which occurs as far north as northern Alaska. These frogs essentially freeze, with up to 60 percent of their body becoming frozen solid for up to eight months of the year! During this time ice crystals form between the muscles and the skin, and large amounts of glucose enter each cell. The outside of each cell freezes but the glucose in each cell prevents the internal parts of the cell from freezing. During this time, metabolism effectively shuts down and the heart completely stops. Once temperatures increase the frog thaws and prepares for breeding. No other known tetrapod is capable of such feats!

At the other end of the spectrum, some frogs tolerate extremely hot environments. The Ryukyu Kajika Frog (*Buergeria japonica*) inhabits hot springs in Taiwan and the Ryukyu Archipelago in Japan. Tadpoles have been reported to occur in pools of 115°F (46.1°C); average human bath temperatures are 95–104°F (35–40°C).

CONSERVATION

Frogs are vital for proper ecological equilibrium in both aquatic and terrestrial environments. Tadpoles eat algae, keeping aquatic environments from being inundated with detrimental algal blooms. Adults eat pest invertebrate species that could otherwise cause devastation to natural environments and result in enormous economic impact upon agriculture. Frogs are food for countless other animals, including terrestrial mammals, bats, birds, reptiles, and invertebrates. Additionally, they help humans beyond being ecologically important by acting as bioindicators and keeping mosquito-borne diseases at bay (some tadpoles eat mosquito larvae and adult frogs eat the adults).

THREAT STATUS

Amphibians are the most threatened terrestrial vertebrate group on the planet, with tens of species likely going extinct each year. The International Union for Conservation of Nature (IUCN) is the primary international organization for the conservation of animals, plants, and fungi. The IUCN maintains the Red List of Threatened Species, which is the main resource for categorizing the threat status of organisms. To date, 6,580 of the 7,600+ known frog species have been evaluated, and only 46 percent of these are considered to be free from any imminent danger of becoming extinct. We know of at least 33 frog species that have become completely extinct in recent years and two that are Extinct in the Wild; however, this number could be much higher, as many species have not been seen for decades. Some estimates suggest that upward of 200 species have gone extinct in recent years. Over 1,500 species are in immediate danger of becoming extinct and are classified as either Endangered or Critically Endangered. Most of the approximately 2,000 species that have been classified as Data Deficient

OPPOSITE | The Kihansi Spray Toad (*Nectophrynoides asperginis*) has been the focus of a successful captive breeding program and reintroductions have started to take place in the wild.

LEFT | The Southern Gastric Brooding Frog (*Rheobatrachus silus*) of Australia is now extinct, possibly caused by chytrid fungus.

or have not been evaluated are almost certainly at risk of extinction because they have not been evaluated due to a general lack of observations, probably because they are rare.

There are numerous threats to amphibian populations, including loss of habitat, disease, climate change, agricultural practices, pollution, and invasive species. Unfortunately, it is usually a combination of these factors that are causing amphibian populations to suffer.

HABITAT LOSS

Habitat loss and the conversion of amphibian habitat into agricultural land is the greatest factor causing declines. Many of the areas with the highest diversity, such as Brazil and Madagascar, have been the most affected by habitat destruction.

Habitat destruction comes in various forms, and the most notable contributions to amphibian declines are total loss of habitat, removal of aquatic breeding grounds, and habitat fragmentation. While fragmentation may not necessarily cause noticeable declines at first, it can lead to a loss of genetic diversity—meaning that the animals become less fit within their environment or are unable to adapt to changing conditions.

The Kihansi Spray Toad (*Nectophrynoides asperginis*) from the Udzungwa Mountains in Tanzania is now considered Extinct in the Wild, according to the IUCN Red List. The species was only known within the spray zone of the Kihansi Falls. A dam constructed farther up the Kihansi River decreased the flow of water passing over the falls by 90 percent, removing the only known habitat of the species. This is just one of hundreds, possibly thousands, of examples of habitat loss that have led to catastrophic declines in frog populations.

DISEASE

Disease is a major cause of frog declines. The most prominent and devastating to frog populations is the amphibian chytrid fungus *Batrachochytrium dendrobatidis*, which is a fungal disease that often leads to the death of its hosts. However, some species seem more resistant than others and the fungus has only limited effects. The fungus grows in the skin layers and can eventually lead to heart failure given that the frog's permeable skin cannot regulate water and gas intake.

Chytrid fungus has caused an estimated 90+ species to go extinct and hundreds of other species to decline significantly. The fungus has been confirmed in every amphibian-inhabited region in the world except for the Seychelles Archipelago. The worst-affected regions are Central and South America, and Australia.

You can help stop the spread of chytrid between populations by maintaining good biosecurity and ensuring footwear is clean and free of dirt when traveling to new areas.

ABOVE | Drought and chytrid fungus (*Batrachochytrium dendrobatidis*) has caused the Northern Corroboree Frog (*Pseudophryne pengilleyi*) to be isolated to just three small subpopulations in southeast Australia.

BELOW | The Cusco Andes Frog (*Bryophryne cophites*) from Peru has been killed by the lethal amphibian chytrid fungus (*Batrachochytrium dendrobatidis*).

CLIMATE CHANGE

Frogs are ectothermic animals with osmotically sensitive physiology, so it is no surprise that their ecologies and distributions are intrinsically linked to climatic variables. It is widely believed that the rapid onset of climate change means that most frog species will be unable to adapt quickly enough to these changes. We are already seeing the results of climate change on frogs from around the world. The Lowland Leopard Frog (*Lithobates yavapaiensis*) from North America has undergone massive declines in many parts of its range due to severe droughts, and the Northern Corroboree Frog (*Pseudophryne pengilleyi*) from Australia has declined and lost at least 42 percent of its breeding sites due to drought.

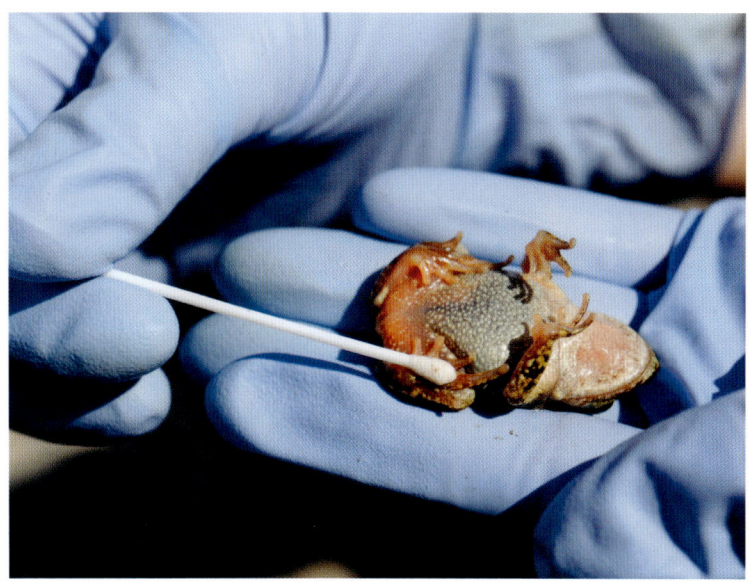

LEFT | A swab sample is being taken from a Majorcan Midwife Toad (*Alytes muletensis*) to test for the presence of the amphibian chytrid fungus (*Batrachochytrium dendrobatidis*).

CHEMICAL CONTAMINATION

Amphibian survival is closely linked to their environment. Frogs have highly sensitive skin important for transfer of gases and water into their body, their eggs are porous, and the larvae have gills. Environmental contaminants from an industrial or agricultural landscape can have cataclysmic effects on amphibians, leading to death, a loss of fitness, or infertility. Chemicals, including antimicrobials and microplastics as well as other common pesticides and heavy metals, usually accompany other factors such as habitat loss, causing increased pressure on populations.

CONSERVATION ACTIONS

It is not all doom and gloom, though, and there have been some fantastic conservation success stories. Conservation actions are driven by dedicated people throughout the world, including academic institutions, conservation organizations, governmental agencies, and members of the public. These inspirational people have at times brought species back from the brink of extinction, whether directly or indirectly, even by doing seemingly simple things like recording sightings of species. There are three major amphibian organizations dedicated to the conservation of amphibians, working very closely with one another but each having unique responsibilities:

1.

The Amphibian Survival Alliance promotes and coordinates conservation actions for amphibians, including outreach with local communities and establishing partnerships.

2.

The IUCN SSC Amphibian Specialist Group provides the scientific foundation to inform effective amphibian conservation action.

3.

The Amphibian Ark leads on the *ex situ* conservation of amphibians, building in-country expertise, and Conservation Needs Assessments (a framework for identifying immediate conservation requirements for individual species).

SUCCESS STORIES

We will finish the introductory section of this book by discussing some of the incredible stories of species recovery. The Mountain Chicken Frog (*Leptodactylus fallax*) is the largest amphibian in the eastern Caribbean, occurring on the small islands of Dominica and Montserrat. The Mountain Chicken has been an economically important species for the people living on the islands, acting as a major food source. In 2002 (Dominica) and 2009 (Montserrat), chytrid devastated the populations, resulting in one of the most rapid declines in a vertebrate species ever recorded. It is likely that no wild individuals now survive on Montserrat. Due to a fast response from the amphibian conservation community, however, Mountain Chickens were taken into captivity to form a successful conservation breeding program in Europe. Subsequently, a breeding facility and a genetics diagnostics lab that performs regular testing for chytrid has been established on Dominica. Most recently a seemingly successful semi-wild control area has been established to investigate reintroduction potential—and the results from this look promising, meaning that the future of the Mountain Chicken might be secured.

The Pool Frog (*Pelophylax lessonae*) is a ubiquitous species across many parts of Europe but native populations became extinct in the UK in 1995 due to a loss of habitat. By translocating animals from Sweden (part of the same clade as the native UK population), two sites in Norfolk in eastern England were selected for a series of reintroductions. These reintroductions have so far been successful, with yearly breeding over the past few years, showing very positive signs for the species in the UK.

OPPOSITE ABOVE | Map showing global distribution of the Anura. Colors denote approximate species diversity with red representing highest diversity areas through to blue, indicating lowest diversity.

OPPOSITE BELOW | It is hoped that with the successful reintroduction program the Pool Frog (*Pelophylax lessonae*) will become more widespread in Britain over the coming years.

LEFT | A long-term recovery strategy has been put in place to help conserve the Mountain Chicken Frog (*Leptodactylus fallax*).

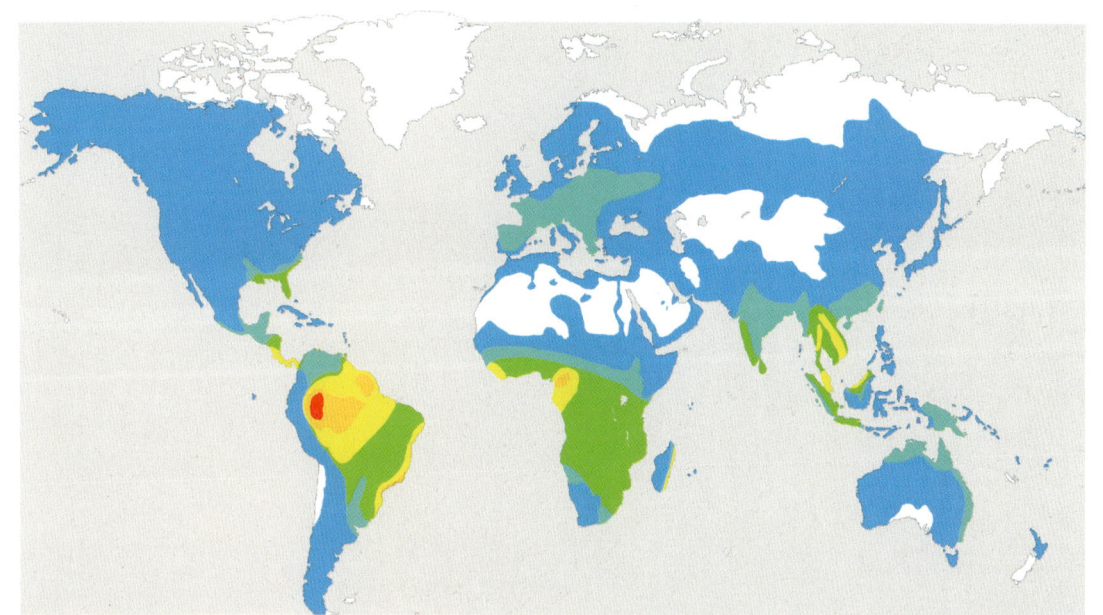

THE FROG SUBORDERS AND SUPERFAMILIES

The Anura contains three suborders, the Archaeobatrachia, Mesobatrachia, and Neobatrachia, although authors have differed with regard to which families comprise the first two of these.

The Archaeobatrachia (archaeo- = ancient; batrachia = frogs) contains the oldest extant frog families, the Ascaphidae and Leiopelmatidae of NW North America and New Zealand respectively, which together comprise the superfamily Leiopelmatoidea, and the European Alytidae and Bombinatoridae, forming the superfamily Discoglossoidea. These extant frogs are linked to long extinct frogs by a number of primitive anatomical characters (see page 59).

The Mesobatrachia (meso- = middle; batrachian) contains those frog families that are more recent than the archaeobatrachians, but which still exhibit primitive anatomical characters (see page 69). This suborder comprises two superfamilies, the Afro-American Pipoidea, with two families, and Eurasian-American Pelobatoidea, with four families.

All other extant anurans belong to the complex Neobatrachia (neo- + new; batrachia). The most basal family is the South African Heleophrynidae, with the remaining families in the Sooglossoidea (2 families, Seychelles and India), Myobatrachoidea (3 families, Australasia and Chile), Hyloidea (21 families), and Ranoidea (19 families).

BELOW | The Strawberry Poison Frog (*Oophaga pumilio*) comes in as many as thirty different color phases ranging from bright red, red with blue legs (blue jeans phase), to green with black spots.

ARCHAEOBATRACHIA

LEIOPELMATOIDEA
Ascaphidae
Leiopelmatidae

DISCOGLOSSOIDEA
Alytidae
Bombinatoridae

MESOBATRACHIA

PIPOIDEA
Pipidae
Rhinophrynidae

PELOBATOIDEA
Pelobatidae
Pelodytidae
Scaphiopodidae
Megaphryidae

NEOBATRACHIA

Heleophrynidae

SOOGLOSSOIDEA
Nasikabatrachidae
Sooglossidae

MYOBATRACHOIDEA
Limnodynastidae
Myobatrachidae
Calyptocephalellidae

HYLOIDEA
Allophrynidae
Centrolenidae
Leptodactylidae
Hylodidae
Aromobatidae
Dendrobatidae
Alsodidae
Cycloramphidae
Odontophrynidae
Rhinodermatidae
Bufonidae
Hylidae
Ceratophryidae
Batrachylidae
Telmatobiidae
Hemiphractidae
Brachycephalidae
Ceuthomantidae
Craugastoridae
Strabomantidae
Eleuthrodactylidae

RANOIDEA
Mantellidae
Rhacophoridae
Pyxicephalidae
Conrauidae
Petropedetidae
Phrynobatrachidae
Ptychadenidae
Ranidae
Ceratobatrachidae
Micrixalidae
Ranixalidae
Nyctobatrachnidae
Dicroglosside
Odontobatrachidae
Hemisotidae
Brevicipitidae
Hyperoliidae
Arthroleptidae
Microhylidae

ARCHAEOBATRACHIA

The Archaeobatrachia contains 4 primitive anuran families, 7 genera, and 27 species, survivors from the dinosaur age. The Ascaphidae of northwestern North America, and the Leiopelmatidae from New Zealand, date back to the Late Jurassic, 163–145 MYA. The Alytidae and Bombinatoridae of Eurasia are traced back to the Mid-Jurassic, 180 MYA, with some of the earliest fossils from England.

These frogs exhibit a suite of primitive characteristics absent in modern anurans. Modern frogs have five to eight presacral vertebrae but the Ascaphidae and Leiopelmatidae possess nine, while Jurassic *Prosalirus bitis* and *Vieraella herbstii* had ten. These extinct Jurassic frogs also had ribs which are retained in Ascaphidae and Leiopelmatidae on the II–IV vertebrae. The Alytidae and Bombinatoridae have eight presacral vertebrae, also with ribs on numbers II–IV. All modern frogs lack ribs.

Ascaphid and leiopelmatid frogs have amphicoelous vertebrae (see Glossary), while alytid and bombinatorid frogs possess opisthocoelous vertebrae, and all modern frogs have procoelous vertebrae. The ascaphids and leiopelmatids also possess an epipubic cartilage anterior to the pubic bone, a feature otherwise only seen in some pipid frogs.

TAILED FROGS

The tailed frogs do not actually possess a tail; rather, males possess a highly vascularized intromittent organ, a cloacal extension that resembles a tail. Supported by cartilaginous strands known as Nobelian rods, this is a ridged structure that is folded forward and inserted into the female's cloaca when the pair are in amplexus. Tailed frogs are the only frogs that internally fertilize their ova.

The Acaphidae is an ancient family from the Late Jurassic (c.163–145 MYA). Today it is represented by one genus and two species, the Coastal Tailed Frog (*Ascaphus truei*) and the Rocky Mountain Tailed Frog (*A. montanus*), which inhabit southern British Columbia, Canada, and parts of five US states (California, Washington, Oregon, Montana, and Idaho). They occur from near to sea level on the coast to 8,390 ft (2,557 m) elevation in the Rocky Mountains. These are delicate frogs that are dorsally brown or gray with translucent pink venters.

Ascaphus is an ancient genus, as evidenced by its nine amphicoelus presacral vertebrae, three pairs

LEFT | Tailed frogs (*Ascaphus* spp.) lack a tympanum (external ear drum).

OPPOSITE | The "tail" of the male Coastal Tailed Frog (*Ascaphus truei*) on the right is clearly visible.

DISTRIBUTION
Pacific northwest United States and southwest Canada

GENUS
Ascaphus

HABITATS
Cold torrential rocky streams through humid forest to 8,390 ft (2,557 m) ASL

SIZE
1¼ in (30 mm) ♂ to 2 in (50 mm) ♀ Coastal Tailed Frog (*A. truei*) and Rocky Mountain Tailed Frog (*A. montanus*)

ACTIVITY
Nocturnal; highly aquatic, occasionally terrestrial

REPRODUCTION
Inguinal amplexus with internal fertilization using male's taillike copulatory organ, eggs laid in bead-like strings attached under stones

DIET
Terrestrial and aquatic insects

IUCN STATUS
None designated

of ribs, and an epipubic cartilage. Tailed frogs also lack a tympanum, the external eardrum visible in most frogs, and males do not call to attract a mate, possibly because the sound of the water might drown out their calls.

Tailed frogs are highly aquatic with extensively webbed toes, but they leave the water to venture into woodlands on wet nights. When they mate, the male grasps the female using inguinal amplexus. Mating occurs in the water and females lay strings of 28–96 eggs attached to the undersides of streambed stones.

Tadpoles inhabit fast-flowing water and exhibit small tail fins and large oral suckers to anchor themselves and graze algae and they have been observed using these suckers to clamber up rocks in the splash zone. Tadpoles may take up to five years to fully develop.

Neither species is considered endangered by the International Union for Conservation of Nature (IUCN), although A. truei is protected in California and Oregon.

NEW ZEALAND FROGS

The Leiopelmatidae contains a single endemic New Zealand genus with four species. Among the most primitive living frog families, it dates back to the Late Jurassic (c.163–145 MYA). Leiopelmatids exhibit nine amphicoelous presacral vertebrae, three pairs of ribs, and an epipubis.

The smallest species is Archey's Frog (*Leiopelma archeyi*). This species inhabits the Coromandel Peninsula and Whareorino Forest on North Island. The largest species is Hamilton's Frog (*L. hamiltoni*), confined to Stephens Island in the Cook Strait. Hochstetter's Frog (*L. hochstetteri*) is the most widely distributed species, distributed through most of the northern half of North Island and Great Barrier Island. The species with the smallest range is the recently described Maud Island Frog (*L. pakeka*) from a single island in the Marlborough Sounds.

New Zealand frogs are dorsally cryptically patterned with brown, gray, or green, and darker below. *Leiopelma hamiltoni* and *L. hochstetteri* have raised dorsal tuberculate ridges on their backs, while *L. archeyi* is smooth-skinned or exhibits only small ridges.

DISTRIBUTION New Zealand	(52 mm) ♀ Hamilton's Frog (*L. hamiltoni*)	limbs, or < 19 eggs laid on damp ground (*L. archeyi* and *L. hamiltoni*) hatching as froglets, without larval stage
GENUS *Leiopelma*	**ACTIVITY** Nocturnal; terrestrial, semiaquatic, occasionally scansoriall	
HABITATS Lowland to upland, and under rocks, logs, or leaf-litter		**DIET** Insects, spiders, and mites
	REPRODUCTION Inguinal amplexus; < 22 eggs laid in small pools (*L. hochstetteri*) hatching into larvae with part-developed	**IUCN STATUS** CR = 1, VU = 2; percentage species in trouble = 75%
SIZE 1¼ in (31 mm) ♂ Archey's Frog (*L. archeyi*) to 2 in		

Inhabiting native forests close to running water, *L. archeyi*, *L. hamiltoni*, and *L. pakeka* adopt a terrestrial lifestyle while *L. hochstetteri* is more semiaquatic with strongly webbed hindfeet. *Leiopelma archeyi* may also climb into low vegetation. They feed on insects and small arachnids but being unable to extend their tongues must lunge forward open-mouthed to catch prey.

Males do not call. *Leiopelma hochstetteri* lays up to 22 eggs in a small wet depression, which hatch into tadpoles with partially developed legs and a tail.

Females of terrestrial species lay up to 19 eggs in leaf-litter where they are guarded by the male. These hatch into tiny froglets with short tails, which ride on their father's back until fully developed.

Leiopelma archeyi is listed as Critically Endangered, while *L. hamiltoni* and *L. pakeka* are Vulnerable. New Zealand frogs are listed in the EDGE of Existence program's 100 Evolutionary Distinct and Globally Endangered amphibians, with *L. archeyi* at the top of the list.

MIDWIFE TOADS & PAINTED FROGS

The Alytidae (formerly Discoglossidae) dates back to the Mid-Jurassic (c.180 MYA) and comprises three Mediterranean genera. They possess the modern number of eight sacral vertebrae, but they are opisthocoelous in arrangement and associated with three pairs of ribs, both primitive characters.

Midwife toads, *Alytes* (6 spp.), inhabit southwestern Europe, northwest Africa, and the Balearic Islands. They are small, stocky frogs with large eyes, and a dorsal pattern of greens, grays, and browns. The most widely distributed is the Common Midwife Toad (*A. obstetricans*) and its name hints at the curious reproductive strategy of these frogs. When the female has laid her strings of eggs, the male will wrap them around his hindlegs and carry them for six weeks before depositing them into water,

TOP LEFT | A male Midwife Toad (*Alytes obstetricans*) may carry strings of eggs from several females for six weeks.

LEFT | The Hula Painted Frog (*Latonia nigriventer*) is a "Lazarus species" from Israel.

OPPOSITE | The Iberian Painted Frog (*Discoglossus galganoi*) is widely distributed from Gibralter to Portugal and the Pyrenees.

DISTRIBUTION
Southern Europe, northern Africa, and Israel

GENERA
Alytes, Discoglossus, Latonia

HABITATS
Most habitats near water, including anthropogenic habitats

SIZE
2 in (47 mm) Moroccan Midwife Toad (*A. maurus*) to 3 in (80 mm) North African Painted Frog (*D. pictus*), Moroccan Painted Frog (*D. scovazzi*), and Iberian Painted Frog (*D. galganoi*)

ACTIVITY
Nocturnal and fossorial (*Alytes*) or both day and night, and terrestrial/aquatic (*Discoglossus* and *Latonia*)

REPRODUCTION
Axillary amplexus; < 1,000 eggs laid in loose clumps in water (*Discoglossus*), or 20–100 eggs laid and

often carrying eggs from several females. Midwife toads like dry, rocky, or wooded hillsides near water, with loose soil for burrowing.

Painted frogs, *Discoglossus* (5 spp.), inhabit Spain, Portugal, southern France, Morocco, Sicily, Sardinia, and Corsica, up to 6,230 ft (1,900 m) above sea level (ASL). A former species, the Hula Painted Frog (*Latonia nigriventer*), inhabits Israel. Last seen in the 1950s, following the draining of the Hula marshes, it was declared extinct in 1996, but it was

rediscovered in 2011, making it a "Lazarus species." Painted frogs are distinctively marked with contrasting stripes or large spots on their pale-brown, warty dorsums. They are more robust than midwife toads, with smaller eyes.

The widest distributed are the Iberian Painted Frog (*D. galganoi*) and the Moroccan Painted Frog (*D. scovazzi*), but the North African Painted Frog (*D. pictus*) is expanding into Europe. They inhabit natural and man-made aquatic habitats, including brackish or stagnant water. Females may lay up to 5,000 eggs, fertilized by several males.

Latonia nigriventer is listed as Critically Endangered, *A. muletensis* and *A. maurus* (Mallorcan Midwife Toad and Moroccan Midwife Toad) are Endangered, *A. dickhilleni* (Betic Midwife Toad) is Vulnerable, and *D. montalentii* (Corsican Painted Frog) is Near Threatened.

carried by the male (*Alytes*)
until close to hatching

DIET
Insects and other invertebrates

IUCN STATUS
CR = 1, EN = 2, VU = 1,
NT = 1; percentage species in
trouble = 42%

FIRE-BELLIED TOADS, BELL TOADS & ASIAN FLAT-HEADED FROGS

The Bombinatoridae is an ancient frog family from the Mid-Jurassic (c.180 MYA), with eight opisthocoelous sacral vertebrae, and three pairs of ribs.

Genus *Bombina* (7 spp.) contains two European species, the Fire-bellied Toad (*B. bombina*) in eastern Europe, and the Yellow-bellied Toad (*B. variegata*) of central and southern Europe. The Oriental Fire-bellied Toad (*B. orientalis*) inhabits eastern China, Russia, and the Korean Peninsula. The other four *Bombina* are usually called bell toads. The widest-distributed is the Large-webbed Bell Toad (*B. maxima*).

These are stout-bodied frogs with short limbs, drab brown or green bodies with large warty tubercles, but they have brightly colored undersides,

LEFT | The European Fire-bellied Toad (*Bombina bombina*) has vivid red markings on its belly that it exposes defensively when it adopts the "unkenreflex" posture.

OPPOSITE | Oriental Fire-bellied Toads (*Bombina orientalis*) occur in Russia, China, and the Korean Peninsula.

OPPOSITE RIGHT | Only found on the islands of Palawan and Busuanga, the Philippine Flat-headed Frog (*Barbourula busuangensis*) is Endangered.

DISTRIBUTION
Europe, Turkey, Russia, China, Korea, Vietnam, Philippines, and Borneo

GENERA
Barbourula, Bombina

HABITATS
Small ponds and lakes, even stagnant water and temporary ponds, in woodland or open habitats (*Bombina*) or fast, shallow streams in rainforest habitats (*Barbourula*)

SIZE
1½ in (40 mm) Apennine Yellow-bellied Toad (*Bo. pachypus*) to 3¼ in (85 mm) Philippine Flat-headed Frog (*Ba. busuangensis*)

ACTIVITY
Primarily diurnal but also active at night

which are displayed when the toad feels threatened and adopts a posture known as "unkenreflex," rolling its hands and feet upside down above its head and displaying the aposematic ventral colors. This is not a bluff, *Bombina* sequester toxins from their invertebrate prey into their own skins.

Males make a soft call by passing air from their vocal sacs to the lungs, the reverse of the way most other frogs call. Females attach their eggs to submerged vegetation.

Genus *Barbourula* (2 spp.), the flat-headed frogs, were originally included with *Alytes* and *Dicroglossus*, but are closer to *Bombina*. The Philippine Flat-headed Frog (*Ba. busuangensis*) inhabits Busuanga and Palawan, while the Bornean Flat-headed Frog (*Ba. kalimantanensis*) is endemic to Indonesian Borneo. Highly aquatic with flattened bodies and heads, they inhabit fast-flowing montane or rainforest streams respectively. Although poorly known, some tantalizing clues have been uncovered. Known from only two specimens, the Bornean Flat-headed Frog is the world's only known lungless frog, respiring by gas exchange through the skin. The reproductive strategy of the Philippine Flat-headed Frog is unknown, but as tadpoles have not been found and females contain very large eggs, it may be a direct breeder without a larval stage.

Barbourula busuangensis is Endangered, *Bombina lichuanensis* and *Bo. fortinuptialis* (Lichuan Bell Toad and Large-spined Bell Toad) are Vulnerable, while *Ba. kalimantanensis* is Near Threatened.

REPRODUCTION
Males grasp females in inguinal amplexus and fertilize the egg clumps that are attached to aquatic vegetation

DIET
Small invertebrates

IUCN STATUS
EN = 2, VU = 2, NT = 1;
percentage species in trouble = 56%

MESOBATRACHIA

The Mesobatrachia contains frogs that are more advanced
than those in the Archaeobatrachia, but most families still
date back to the Late Jurassic or Early Cretaceous, 163–100
MYA, and they exhibit some primitive characteristics such
as amphicoelous or opisthocoelous vertebrae, while the
vertebrae of modern, neobatrachian frogs are procoelous.
The Mesobatrachia contains six families split between two
superfamilies (suffix = -oidea).

The Pipoidea contains the entirely aquatic Pipidae, which
is divided into two subfamilies, the South American Pipinae
and sub-Saharan African Dactylethrinae. Also included in
the Pipoidea is the monotypic, fossorial Rhinophrynidae
from Mexico.

The Pelobatoidea comprises the Pelobatidae from Europe, Asia,
and North Africa, the Eurasian Pelodytidae, the North American
Scaphiopodidae, and the large Asian Megophrynidae. Most
members of the Pelobatoidea are terrestrial frogs, in contrast to
the aquatic and fossorial Pipoidea. The Pelobatidae and
Pelodytidae diverged around 150 MYA.

The Mesobatrachia contains 360 species in 20 genera, a small
proportion of the total number of 7,600+ frogs described.

AFRICAN CLAWED FROGS & PLATANNAS

Dactylethrinae contains three genera of sub-Saharan African aquatic clawed frogs. The largest is *Xenopus* (29 spp.), although some authors place the Tropical Clawed Frog (*X. tropicalis*) and three rainforest *Xenopus* in *Silurana*. They are called "platannas" in South Africa, in Afrikaans meaning "flat-handed." *Hymenochirus* (4 spp.) are dwarf clawed frogs from the West and Central African rainforests, the largest being the Gaboon Dwarf Clawed Frog (*H. feae*), while Merlin's Dwarf Clawed Frog (*Pseudohymenochirus merlini*) is a monotypic species from Guinea-Bissau and Sierra Leone. Because of its importance to science (see page 71), the Smooth Clawed Frog (*Xenopus laevis*) was the first amphibian to have its entire genome mapped.

Clawed frogs are dorsoventrally flattened and pear-drop-shaped, with smooth skin, large dorsally positioned eyes, wide splayed legs, webbed toes, and long clawed fingers. *Xenopus* have nictitating membranes over their eyes and an epipubis on the pelvis, features absent in the other two genera. Although fully aquatic, they can crawl over land.

Xenopus tadpoles filter-feed but those of the other genera are carnivorous. Adults feed on aquatic invertebrates, fish, tadpoles, and smaller frogs, but large species can take birds or small mammals.

DISTRIBUTION
Sub-Saharan Africa, including Bioko Island, and introduced worldwide

GENERA
Hymenochirus, Pseudohymenochirus, Xenopus

HABITATS
Still or slow-moving waters, from lakes to stagnant pools and man-made fish ponds, in close canopy rainforest to open grassland, and perianthropic habitats

SIZE
1 in (27 mm) Boulenger's Dwarf Clawed Frog (*H. boulengeri*) to 5 in (130 mm) ♀ Smooth Clawed Frog (*X. laevis*)

ACTIVITY
Nocturnal, diurnal; fully aquatic

REPRODUCTION
Males grasp females in inguinal amplexus and fertilize the eggs, which are attached to aquatic vegetation, usually at night

ABOVE | Boettger's Dwarf Clawed Frog (*Hymenochirus boettgeri*) is a rainforest species from southern Nigeria to the Democratic Republic of the Congo.

OPPOSITE TOP | The Lake Oku Clawed Frog (*Xenopus longipes*) is confined to one Cameroonian lake and is Critically Endangered.

OPPOSITE | Once used globally as an economical means of pregnancy testing, the Smooth Clawed Frog (*Xenopus laevis*) may have been the vehicle for the global transmission of chytridiomycosis.

DIET
Aquatic invertebrates, tadpoles and smaller frogs, fish, and even birds or small mammals

IUCN STATUS
CR = 2, EN = 3, VU = 1, NT = 1; percentage species in trouble = 18%

Small prey is just swallowed but larger prey is held in the mouth and ripped apart using the clawed fingers. To avoid predation, they possess skin peptides that induce vomiting, resulting in avoidance therapy that protects the general population.

Males court females by making underwater clicking sounds. Adult *Hymenochirus* and *Pseudohymenochirus* perform aquatic somersaults as the male fertilizes the female's eggs. Male *Pseudohymenochirus* flick the eggs onto the female's back like Suriname toads (*Pipa*, see page 72), while *Xenopus* and *Hymenochirus* lay theirs at the surface.

African clawed frogs are believed to have been the vehicle for the global transmission of the deadly amphibian fungal disease chytridiomycosis, but some *Xenopus* are under threat too. *Xenopus longipes* and *X. lenduensis* (Lake Oku Clawed Frog and Lendu Clawed Frog) are Critically Endangered, three species are Endangered, and one each Vulnerable and Near Threatened.

SURINAME TOADS

ABOVE | The Arrabal's Suriname Toad (*Pipa arrabali*), from Guyana, is one of the species that produces fully formed froglets rather than tadpoles. Here is a female and a newborn froglet.

OPPOSITE | A female Amazonian Suriname Toad (*Pipa pipa*), with newly laid and fertilized eggs on her back, is a very strange sight and testament to the diverse reproductive histories of amphibians.

The Pipinae comprises genus *Pipa* (7 spp.), the Suriname toads. The most familiar species is the Amazonian Suriname Toad (*P. pipa*), but the genus occurs from Panama (Myers' Suriname Toad, *P. myersi*) to Atlantic coastal Brazil (Carvalho's Suriname Toad, *P. carvalhoi*). Recent molecular studies suggest there are cryptic species awaiting discovery.

These are dorsoventrally compressed frogs with dark, pear-shaped bodies; angular, pointed heads with dorsolateral eyes; widely splayed limbs; and strongly webbed toes. All species lack tongues and some lack teeth. The largest species, the Amazonian Suriname Toad and the Utinga Suriname Toad (*P. snethlageae*), also Amazonian, possess loose flaps of skin at the corners of their mouths.

Suriname toads have long fingers bearing terminal lobes with tripartite or quadripartite, bifurcated or non-bifurcated tips, characteristics used in species identification. These lobes are sensory organs for prey location.

Suriname toads exhibit one of the strangest anuran reproductive strategies. Amplexus may last 12 hours, during which time the male will perform aquatic somersaults as he fertilizes the female's eggs, before flicking them onto the female's back where they sink into depressions and become

DISTRIBUTION
Panama and northern South America

GENUS
Pipa

HABITATS
Slow-moving rivers with muddy bottoms, marshes, lakes, and ponds

SIZE
1¾ in (44 mm) ♂ Sabana Suriname Toad (*P. parva*) to 6¾ in (171 mm) ♀ Amazonian Suriname Toad (*P. pipa*)

ACTIVITY
Nocturnal or diurnal

REPRODUCTION
Inguinal amplexus and somersaulting for up to 12 hours, the eggs laid in the skin of the female's back where they incubate to become free-swimming tadpoles or fully developed froglets, depending on species

absorbed by the rapidly developing skin to begin incubation. After a few weeks the eggs hatch under the skin. In three species, *P. myersi*, *P. carvalhoi*, and the smallest species, the Sabana Suriname Toad (*P. parva*) of Colombia and Venezuela, tadpoles emerge from the female's back and escape into the water to begin a free-swimming larval stage.

In the remaining four species, Arrabal's Suriname Toad (*P. arrabali*) and Rough Suriname Toad (*P. aspera*) from the Guianas, and the Amazonian Suriname Toad and Utinga Suriname Toad, females continue to incubate their eggs for three to four months, after which tiny froglets will haul themselves out of craters on their mother's backs. The elimination of the free-swimming larval stage is called "direct development" and many unrelated frogs practice it, though not in the same style.

Only *P. myersi* (Myers' Suriname Toad) is listed as Endangered.

DIET
Aquatic invertebrates, earthworms, and small fish

IUCN STATUS
EN = 1, percentage species in trouble = 14%

MESOAMERICAN BURROWING TOAD

Although the sole member of the Rhinophrynidae is usually called the Mexican Burrowing Toad (*Rhinophrynus dorsalis*), a more appropriate common name is Mesoamerican Burrowing Toad, because it is found in subtropical and tropical habitats from southern Texas down the Caribbean versant to the Yucatan Peninsula, Belize, and western Honduras, and down the Pacific versant from Jalisco, Mexico, to Costa Rica. This is wider distribution than "Mexican" would suggest.

It has a globular pear shape, with a short, pointed, reinforced head, small protruding dorsolateral eyes, no teeth or tympanum,

DISTRIBUTION
Southern Texas to Costa Rica

GENUS
Rhinophrynus

HABITATS
Subtropical and tropical forest and agricultural habitats

SIZE
3¼ in (85 mm) Mesoamerican Burrowing Toad (*R. dorsalis*)

ACTIVITY
Primarily fossorial except after rain

REPRODUCTION
Inguinal amplexus; an explosive breeder in temporary pools with females laying

thousands of eggs that sink to the bottom

DIET
Soft-bodied subterranean invertebrates, e.g., termites and ants

IUCN STATUS
Low Concern, no species are listed as under threat

and short but powerful limbs, the toes being strongly webbed. Coloration is black with a distinctive orange vertebral stripe and orange spots and blotches on the flanks and limbs. The feeding mechanics of the mouth are modified for an entirely myrmecophagous diet, the tongue passing out of the front of the mouth through a groove.

Although the Mesoamerican Burrowing Toad bears a very strong resemblance to the purple pig-nosed frogs (*Nasikabatrachus*, see page 93) of southern India, and the Turtle Frog (*Myobatrachus*, see page 96) of Australia, the three genera are not related; they are examples of convergence having adopted a similar, highly fossorial existence, feeding on soft-bodied termites and ants, only emerging onto the surface after heavy rain, which is when *Rhinophrynus* engages in explosive breeding in temporary ponds and puddles.

Males float and call on the surface before clasping the females, which lay thousands of eggs that sink to the bottom of the pond. The tadpoles begin as filter-feeders but rapidly develop carnivorous mouthparts and may even become cannibalistic. They form shoals of 50–100 tadpoles for protection.

It has been suggested that the Mesoamerican Burrowing Toad is separated from all other anurans by at least 190 million years of evolution and may be the most evolutionarily distinct living amphibian in the world. It is listed as Low Concern, and not believed to be under threat.

ABOVE | Mesoamerican Burrowing Toads (*Rhinophrynus dorsalis*) only emerge onto the surface after heavy rain when they breed in shallow ponds.

OPPOSITE | The pear-shaped Mesoamerican Burrowing Toad is adapted to feed on ants and termites.

NEARCTIC SPADEFOOT TOADS

LEFT | Couch's Spadefoot Toad (*Scaphiopus couchii*) will make use of ephemeral man-made pools and ponds if no natural watercourses are available.

OPPOSITE | The Eastern Spadefoot Toad (*Scaphiopus holbrooki*) has been recorded using the same burrow for up to five years.

OPPOSITE BELOW | Affected by agriculture and urbanization, the Western Spadefoot Toad (*Spea hammondi*) is listed as Near Threatened by the IUCN.

The spadefoot toads of the USA, Canada, and Mexico belong to two genera, *Scaphiopus* (3 spp.) in the eastern states, and *Spea* (4 spp.) in the western states, with an overlap of both genera in the southern states. The spades are raised black tubercles on the underside of the hindfeet. Used for digging backward, they served to distinguish the two genera, being wedge-shaped in *Spea* and sickle-shape in *Scaphiopus*. In *Scaphiopus* the eyelids are the same width as the space between them on the dorsum of the head, while in *Spea* the eyelids are wider than the gap between them.

Nearctic spadefoot toads are short, chunky toads with smooth pastel-colored skin, squarish snouts, large eyes with vertical pupils, and short legs, for digging rather than leaping.

The Eastern Spadefoot (*Scaphiopus holbrookii*) occurs from New England to Florida, and west to Tennessee and Louisiana. In Louisiana, Arkansas, and Texas, it gives way to Hurter's Spadefoot (*Sc. hurterii*), and then Couch's Spadefoot (*Sc. couchii*), which is distributed across the southwest to California, and south to Baja California and Veracruz, Mexico. The Mexican Spadefoot

DISTRIBUTION
North America and Mexico

GENERA
Scaphiopus, Spea

HABITATS
Desert, semidesert, prairie grassland, chaparral, with loose soils and close to pools, ditches or streams, also some agricultural but not urban areas

SIZE
2½ in (63 mm) Plains Spadefoot (*Sp. bombifrons*) to 3½ in (90 mm) Couch's Spadefoot (*Sc. couchii*)

ACTIVITY
Nocturnal; terrestrial, semi-fossorial, aquatic

REPRODUCTION
Explosive breeders following heavy rain, males jostling for

females, adopting inguinal amplexus, with rapid development to metamorphosis (8–16 days)

DIET
Small invertebrates

IUCN STATUS
NT = 1; percentage species in trouble = 14%

(*Spea multiplicata*) occurs from Oaxaca, north to Utah and Colorado, overlapping the range of the Plains Spadefoot (*Sp. bombifrons*), which is distributed north to Alberta and Saskatchewan, Canada. To the west, the Great Basin Spadefoot (*Sp. intermontana*) occurs from Nevada to British Columbia, with the Western Spadefoot (*Sp. hammondi*) inhabiting the coastal strip from southern California to northern Baja California.

Spadefoots like open habitats, for example, desert, semidesert, prairie, or chaparral, and prefer loose soil for digging. After heavy spring and summer rains males emerge from burrows to call for females. Eggs laid in temporary pools develop rapidly through the larval stage to metamorphosis before the pools dry out. Spadefoots spend the drier months estivating in animal burrows or in self-dug excavations.

Spea hammondi is Near Threatened, while several other species receive state protection or are listed as Species of Special Concern within Canadian or US jurisdictions.

PALEARCTIC SPADEFOOT TOADS

LEFT | The Common Spadefoot Toad (*Pelobates fuscus*) of eastern and central Europe will defend itself by secreting a garlic-like toxin and attempting to bite its enemy.

OPPOSITE | Once widely distributed from Turkey to Iran and Jordan, the range of the Syrian Spadefoot Toad (*Pelobates syriacus*) is now much smaller.

OPPOSITE BELOW | The Moroccan Spadefoot Toad (*Pelobates varaldii)* is only known from the northwest Atlantic coast of Morocco, where it is Endangered.

The Pelobatidae contains just genus *Pelobates* (6 spp.), which is distributed across Europe and Asia as far as Turkey, the Caucasus, Iran, and Kazakhstan. The Western Spadefoot Toad (*P. cultripes*) inhabits the Iberian Peninsula and southern France, the Balkan Spadefoot Toad (*P. balcanicus*) is found in the Balkans and Greece, and the Common Spadefoot Toad (*P. fuscus*) occurs in northern Italy and central Europe, eastward to Russia and Kazakhstan where it meets with the range of the Russian Spadefoot Toad

(*P. vespertinus*). Turkey, Jordan, Lebanon, Israel, Syria, Iran, and the Caucasus are inhabited by the Syrian Spadefoot Toad (*P. syriacus*), while the Moroccan Spadefoot Toad (*P. varaldii*) is found in northwest Africa.

Spadefoot toads have squat bodies, short limbs with webbed toes, and large eyes with vertical pupils. Most spadefoot toads are a pale gray, green, or brown above with a series of large irregular darker blotches. Their name "spadefoot" comes from a large metatarsal tubercle on their hindfeet

DISTRIBUTION
Europe, western Asia, and northwest Africa

GENUS
Pelobates

HABITATS
Open habitats, e.g., sand dunes, grassland, agricultural land, and a wide variety of watercourses

SIZE
2½ in (60 mm) Moroccan Spadefoot Toad (*P. varaldii*) to 4 in (100 mm) European Spadefoot Toad (*P. cultripes*)

ACTIVITY
Nocturnal; fossorial, aquatic

REPRODUCTION
Explosive breeders, males adopting inguinal amplexus,

females laying thousands of eggs in strings

DIET
Small invertebrates

IUCN STATUS
EN = 1, VU = 1; percentage species in trouble = 33%

that they use for digging backward into soft soil or sand. On land they occur in open habitats, such as sand dunes or grasslands, where they are extremely fossorial, only emerging at night after rain.

Spadefoot toads are explosive breeders in the spring, being found in a wide variety of watercourses, the males calling the females underwater despite lacking a vocal sac. Females lay several thousand eggs in strings. While the tadpole stage is rapid (two to three months) in the Syrian Spadefoot Toad it is much longer in more northern species such as the Common Spadefoot Toad, where tadpoles may overwinter for up to three years. Tadpoles may be 7 in (180 mm) in length, the longest European anuran larvae. *Pelobates veraldii* is Endangered, *P. cultripes* is Vulnerable, and *P. syriacus* may be extirpated from Jordan and Syria.

PARSLEY FROGS

The Pelodytidae diverged from the Pelobatidae during the Late Jurassic 150 MYA. It contains a single genus, *Pelodytes* (5 spp.) the parsley frogs, two of which were only recently described. The genus exhibits a discontinuous distribution across southwestern Europe to the Caucasus. The Common Parsley Frog (*P. punctatus*) inhabits France, the Low Countries, and northern Italy, while the Iberian Parsley Frog (*P. ibericus*) is found in Spain and Portugal, where there are also two endemics, the Hesperides' (Western) Parsley Frog (*P. hespericus*) in Spain and the Lusitanian Parsley Frog (*P. atlanticus*) in Portugal. Almost 1,860 miles (3,000 km) to the east,

LEFT | The Common Parsley Frog (*Pelodytes punctatus*) can scale smooth vertical surfaces.

OPPOSITE | The Iberian Parsley Frog (*Pelodytes ibericus*) from the extreme south of the Spain and Portugal, was only described in 2000.

DISTRIBUTION
Western Europe and Caucasus

GENUS
Pelodytes

HABITATS
Woodland and grassland on limestone or sandy soils, and a wide variety of watercourses

SIZE
1½ in (39.5 mm) ♂ Iberian Parsley Frogs (*P. ibericus*) to 4 in (100 mm) ♀ Caucasian Parsley Frog (*P. caucasicus*)

ACTIVITY
Nocturnal; terrestrial, aquatic

REPRODUCTION
Males adopt inguinal amplexus; may reproduce more than once a year;

females lay up to 1,600 eggs a year in strings

DIET
Small invertebrates

IUCN STATUS
NT = 1; percentage species in trouble = 20%

the Caucasian Parsley Frog (*P. caucasicus*) inhabits the coasts of the Black and Caspian Seas.

In body shape parsley frogs are typically frog-like with gracile bodies covered with rows of tubercles, pointed snouts, large eyes with vertical pupils, and powerful legs, the toes lacking webbing. They are gray, brown, or green in color with darker spotting.

Parsley frogs are active primarily at night, although they may also be diurnal during the breeding season. They inhabit woodlands or grasslands on limestone or sandy soils, up to 7,545 ft (2,300 m) ASL, and they frequent both shallow and deep waters, from temporary ponds to quarries, ditches, or cisterns. Their legs are powerful for leaping but during the day they usually shelter underneath rocks or logs.

They are active all year around and in the right conditions may reproduce more than once in a single year. Females lay strings of up to 360 eggs, which are attached to aquatic vegetation. In a season a female Common Parsley Frog may lay up to 1,600 eggs, but females of the Caucasian Parsley Frog produce only around 500, which exhibit a rapid developmental period, hatching in five days and metamorphosing into froglets in a month. This species is also considered Near Threatened by the IUCN.

ASIAN LITTER, SPADEFOOT, MUSTACHE, TOOTHED & ALPINE TOADS

The Leptobrachiinae inhabits mainland Asia and Southeast Asia, including the Sunda Islands and the Philippines, and contains over 170 species in 4 genera.

Leptobrachella (91 spp.) contains numerous small cryptic dwarf litter toads distributed from China to Borneo and Natuna Islands. They are forest floor dwellers that occur in a variety of shapes, from slender and gracile to stocky and angular-bodied, and they are usually associated with streams. *Leptobrachium* (38 spp.) contains the Asian spadefoot toads and mustache toads. These are usually larger than the toads in *Leptobrachella*, several species achieving almost 4 in (100 mm) SVL, with stout bodies, short limbs, and often very large eyes, which are pale blue in the Bompu Toad

LEFT | From the Eaglenest Wildlife Sanctuary in northeast India, the Bompu Toad (*Leptobrachium bompu*) has stunning sky-blue eyes.

OPPOSITE | Montane Large-eyed Litter Frogs (*Leptobrachium montanum*), from Borneo, sit bolt upright balanced on their knuckles.

OPPOSITE BELOW | Mjøberg's Dwarf Litter Frog (*Leptobrachella mjobergi*) is a tiny frog from the leaf-litter of Sarawak, Borneo.

DISTRIBUTION
Southern China, northern India, Nepal, Southeast Asia, Greater Sunda Islands, and southern Philippines

GENERA
Leptobrachella, Leptobrachium, Oreolalax, Scutiger

HABITATS
Rainforest, temperate forest, hill forest, mountains, grassland, rocky outcrops, agricultural areas, and near water

SIZE
¾ in (17 mm) Webfoot Dwarf Litter Toad (*Leptobrachella palmata*) to 3¾ in (95 mm) Lowland Spadefoot Toad (*Leptobrachium abbotti*)

ACTIVITY
Nocturnal; terrestrial, semi-fossorial, secretive

REPRODUCTION
Males are often armored with sharp spines during the breeding season, amplexus is inguinal, and females lay their eggs into forest streams, the larvae being free-swimming

(*Leptobrachium bompu*) from northern India and the Way Sepunti Toad (*Leptobrachium waysepuntiense*) of Sumatra, but two-tone in other species. Many are leaf-litter dwellers, but others inhabit rocky crevices. One of the strangest is the Emei Mustache Toad (*Leptobrachium boringii*), males of which are adorned with a series of temporary sharp black spines along the upper lip, the "mustache," but only during the breeding season.

Oreolalax (19 spp.) are the Asian toothed toads, from southwestern China, with one Indo-Chinese species, Sterling's Toothed Toad (*O. sterlingae*). Many *Oreolalax* are also armed with sharp spines during the breeding season, on the dorsum, venter, and forelimbs. The genus *Scutiger* (24 spp.) contains the alpine toads, aka cat-eyed or lazy toads, which are Himalayan in distribution and occur over a greater elevational range (3,280–17,390 ft [1,000–5,300 m]) than any other amphibians, in rocky grassland and fast-flowing streams, with the Xizang Alpine Toad (*S. boulengeri*) inhabiting open alpine desert of the Tibetan plateau, in streams of glacial run-off.

The tadpoles of leptobrachiines show considerable morphological variation, those of the genus *Leptobrachella* being vermiform, while, due to the cold, the tadpoles of *S. boulengeri* take up to five years to metamorphose. Seven species are Critically Endangered.

DIET
Leptobrachella, small invertebrates; *Leptobrachium*, large invertebrates

IUCN STATUS
CR = 7, EN = 36, VU = 13, NT = 7; percentage species in trouble = 37%

ASIAN HORNED TOADS

RIGHT | Annam Horned Toads (*Brachytarsophrys intermedia*) inhabit the highlands of Vietnam and Laos above 2,950 ft (900 m) elevation.

OPPOSITE | The Bornean Horned Toad (*Pelobatrachus nasutus*) is named for the protrusion on its snout.

The Megophryinae inhabits mainland Asia and Southeast Asia and comprises 10 genera and almost 130 species generally called Asian horned toads because they exhibit large raised, forward-facing, supraocular horns that shield their large eyes and aid the crypsis of these mottled brown forest floor anurans. Some species are smooth-skinned while others have rugose dorsums covered in tubercles that further break up their outlines, and most are various shades of reddish-brown.

Camouflage is essential because these toads are heavy-bodied with short limbs and cannot easily escape predators, so instead they vanish in plain sight. The large, broad heads of the megophryines, often equal to half of their SVL, also afford them large mouths for devouring large rainforest invertebrates including scorpions and centipedes, and snails with 2 in (50 mm) shells.

Three genera—*Atympanophrys* (4 spp.); *Boulenophrys* (65 spp.), the Chinese horned toads;

DISTRIBUTION
China, northern India, Nepal, Southeast Asia, Greater Sunda Islands, and Philippines

GENERA
Atympanophrys, Boulenophrys, Brachytarsophrys, Megophrys, Ophryophryne, Pelobatrachus, Xenophrys

HABITATS
Rainforest, temperate forest, hill forest, and rainforest streams

SIZE
1½ in (35 mm) Yu Shen Horned Toad (*Bo. ombrophila*) to 4¾ in (120 mm) Bornean Horned Toad (*P. nasutus*)

ACTIVITY
Nocturnal; terrestrial, semi-fossorial, secretive

REPRODUCTION
Inguinal amplexus

DIET
Large invertebrates including cockroaches, scorpions, centipedes, and even snails

IUCN STATUS
CR = 4, EN = 9, VU = 6, NT = 6; percentage species in trouble = 20%

and *Brachytarsophrys* (8 spp.), the Asian short-legged toads—are primarily Chinese but with several species in Southeast Asia. The strange-horned toads, genus *Xenophrys* (30 spp.), range from Xizang (Tibet), Nepal, and Bhutan, south to Peninsular Malaysia, while the narrow-mouthed horned toads, *Ophryophryne* (7 spp.), occur from southern China to northern Thailand.

The remaining two genera exhibit tropical distributions. The genus *Megophrys* (5 spp.) inhabits Sumatra and Java and includes the Montane Horned Toad (*M. montana*). *Pelobatrachus* (7 spp.) contains those species formerly placed in *Megophrys* from the Malaysian Peninsula, Borneo, and the Philippines, including the most impressive species in

the family, the Bornean Horned Toad (*P. nasutus*). In this species the leaf mimicry is striking, the supraocular horns being complemented by a fleshy nasal projection on the front of the snout. Found throughout the rainforests of Borneo, Sumatra, and the Malay Peninsula, female Bornean Horned Toads achieve 4¼– 4¾ in (110–120 mm) SVL. The loud honking call of the males is often the prelude to heavy rain as they laboriously make their way to streams to breed.

Four species are Critically Endangered with a further nine species Endangered.

NEOBATRACHIA

The Neobatrachia contains all the anurans that are neither in the Archaeobatrachia or the Mesobatrachia, which means 46 families, 430 genera, and almost 7,000 species. The southern African family Heleophrynidae is considered the sister taxon to all other neobatrachians, which are then divided between four superfamilies.

The Sooglossoidea is a small superfamily containing the south Indian Nasikabatrachidae and the Seychelles Sooglossidae, two ancient families originating in Jurassic Gondwanaland, while the Myobatrachoidea, also Gondwanan in origin, contains the Australopapuan Myobatrachidae and Limnodynastidae and the Chilean Calyptocephalellidae.

The remaining 40 families are divided between two huge superfamilies that diverged 170 MYA. The Hyloidea contains 21 families and is primarily a Latin American clade, although it includes the widely distributed Hylidae and Bufonidae. The monophyly of the Hyloidea is supported by molecular data.

The Ranoidea contains 19 families, 11 from Africa or Madagascar, 3 South Asian, 2 Afro-Asian, 1 Southeast Asian, and 2 globally distributed families, the Ranidae and Microhylidae. The monophyly of the Ranoidea is supported by both molecular and morphological data.

GHOST FROGS

Heleophrynidae is a small southern African family of two genera, the sister taxon to all other Neobatrachia. Heleophrynids are called ghost frogs, a name originating from Skeleton Gorge on Table Mountain where *Heleophryne rosei* occurs.

Ghost frogs are small with flattened bodies, narrowing toward the head, which ends in a flattened, rounded snout. The eyes are large and bulbous with vertical pupils, and the limbs are powerful, with variable webbing on the toes, males having more webbing than females, and long fingers and toes

ABOVE | Due to the establishment of pine plantations within its tiny range, the Table Mountain Ghost Frog (*Heleophryne rosei*) is Critically Endangered.

OPPOSITE TOP | The Natal Ghost Frog (*Hadomophryne natalensis*) is the largest ghost frog and the most widely distributed in southeastern Africa.

OPPOSITE INSET | Cape Ghost Frogs (*Heleophryne purcelli*) inhabit the montane fynbos heathlands of the Cederberg range.

DISTRIBUTION
South Africa, Lesotho, and Eswatini

GENERA
Hadromophryne, Heleophryne

HABITATS
Montane forest, grassland and fynbos, and gorges and fast-flowing streams

SIZE
1¾ in (46 mm) ♂ Cederberg Ghost Frog (*He. depressa*) to 2½ in (65 mm) ♀ Natal Ghost Frog (*Ha. natalensis*).

ACTIVITY
Nocturnal; highly aquatic, and terrestrial on rocks

REPRODUCTION
Inguinal amplexus preceded by tactile displays by both sexes,

100–200 eggs laid under rocks in shallow ponds, tadpoles with large suckers for anchorage in fast water; metamorphosis takes 1–2 years

DIET
Presumed small invertebrates

IUCN STATUS
CR = 1, EN = 1; percentage species in trouble = 29%

with triangular discs that afford the frogs excellent grip, all characteristics perfected for life on the rocks in fast-flowing torrents. Coloration is primarily green or brown with distinctive darker blotches and spots.

They inhabit montane torrents in forest, grassland, or fynbos habitats that experience high rainfall. The largest species, the Natal Ghost Frog (*Hadromophryne natalensis*), is found widely across eastern South Africa, Lesotho, and Eswatini. *Heleophryne* (6 spp.) are confined to the Western Cape, a region of high endemicity. They are distributed in an arc from west to east: Cederberg Ghost Frog (*H. depressa*), Cape Ghost Frog (*H. purcelli*), Table Mountain Ghost Frog (*H. rosei*), Eastern Ghost Frog (*H. orientalis*), Southern Ghost Frog (*H. regis*), and Hewitt's Ghost Frog (*H. hewitti*).

Courtship has only been documented in *H. purcelli*, where the smaller males develop loose folds of skin and webbing on their body and arms, and spines on the surfaces that will contact the females. Males call from the water for females and the pairs engage in a period of mutual touching of head and arms before the females lay their eggs under streambed rocks in slower parts of the stream. Tadpoles possess large suctorial discs to maintain anchorage on rocks in fast-flowing water. Development is slow, with tadpoles not metamorphosing into frogs for one to two years, due to the cooler temperatures at higher elevations.

Heleophryne rosei is Critically Endangered while *H. hewitti* is Endangered, due to urban development.

PURPLE PIG-NOSED FROGS

BELOW | The Purple Pig-nosed Frog (*Nasikabatrachus sahyadrensis*) is a fossorial species that engages in explosive breeding events following the rains, males grasping the much larger females in amplexus.

Nasikabatrachidae contains a single genus, *Nasikabatrachus*, and two highly fossorial species. *Nasikabatrachus sahyadrensis*, the Purple Pig-nosed Frog, was only described in 2003, although it was already familiar to local farmers. It is a strange-looking amphibian with a bloated purple body, smooth skin, short limbs, a small, pointed head with small, forward-facing eyes and a proboscis-like snout.

A second species, Bhupathy's Purple Pig-nosed Frog (*N. bhupathi*), was named in 2017 for Subramaniam Bhupathy, an Indian zoologist who died while engaged in fieldwork in 2014. The Purple Pig-nosed Frog inhabits the Western Ghats in Karnataka, southern India, an ancient mountain range demonstrating a huge degree of endemicity across zoology. Bhupathy's Purple Pig-nosed Frog was discovered in an agricultural habitat close to a wildlife sanctuary on the eastern slopes of the Western Ghats, in the southern Provinces of Kerala and Tamil Nadu.

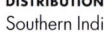

DISTRIBUTION
Southern India (Western Ghats)

GENUS
Nasikabatrachus

HABITATS
Montane rainforest and agricultural fields close to streams

SIZE
2 in (49 mm) Bhupathy's Purple Pig-nosed Frog (*N. bhupathi*) to 3½ in (90 mm) Purple Pig-nosed Frog (*N. sahyadrensis*)

ACTIVITY
Nocturnal; fossorial, secretive

REPRODUCTION
An explosive breeder after heavy rain, males engaging in inguinal amplexus

DIET
Termites

IUCN STATUS
EN = 1; percentage species in trouble = 50%

Morphologically *Nasikabatrachus* resemble the equally fossorial Mesoamerican Burrowing Toad (*Rhinophrynus dorsalis*) and Western Australian Turtle Frog (*Myobatrachus gouldii*). The hands and feet of Bhupathy's Purple Pig-nosed Frog are toughened, and the feet bear enlarged metatarsals for burrowing, while the ossified skull could also be an adaptation for burrowing. The white proboscis-like snout is believed to enable the frog to penetrate the walls of termitaria to lap up termites with its tongue.

The closest relatives of these frogs are the Seychelles frogs of the Sooglossidae. Both families are Gondwanan in origin, dating back to the Jurassic 170 MYA, but diverging when Madagascar, the Seychelles, and India broke away from Gondwanaland during the Cretaceous 130 MYA. The purple pig-nosed frogs only emerge onto the surface to breed following monsoonal rains. Due to their distributions on opposite sides of the Western Ghats and the timing of the monsoons, they breed at different times of year.

Nasikabatrachus sahyadrensis is Endangered and decreasing due to loss of habitat to coffee farming. The status of *N. bhupathi* has not been accessed but its known range is smaller than that of its congeneric, so it is in danger of extinction.

TOP | Tadpoles of the Purple Pig-nosed Frog (*Nasikabatrachus sahyadrensis*) have large suckers to enable them to attach to rocks in fast-flowing torrents.

ABOVE | Bhupathy's Purple Pig-nosed Frog (*Nasikabatrachus bhupathi*) was named in honor of India's leading python expert, Subramanian Bhupathy, who died while conducting fieldwork.

SEYCHELLES FROGS

The Sooglossidae is endemic to the Seychelles, where it only inhabits the larger granitic islands. There are four species in two genera. The Seychelles Frog (*Sooglossus sechellensis*) inhabits Mahé, Silhouette, Praslin, and Cerf, a small island off Mahé, at elevations of 820–3,115 ft (250–950 m) ASL. The smallest species, Gardiner's Frog

(*Sechellophryne gardineri*), which achieves only ½ in (11 mm) SVL, is found on Mahé and Silhouette at 655–2,950 ft (200–900 m) ASL, as is the largest species, at 2¼ in (55 mm), Thomasset's Frog (*So. thomasseti*), although this species is reported over a greater altitudinal range (260–3,250 ft [80–990 m] ASL) than the other species. All three species are nocturnal, terrestrial rainforest leaf-litter dwellers that occur close to fast-flowing streams feeding on small invertebrates.

The Seychelles Palm Frog (*Se. pipilodryas*) is the most arboreal species, inhabiting the axils of the endemic Seychelles Palm (*Phoenicophorium borsigianum*). It is also the most recently described (2002) and probably the most geographically restricted sooglossid, being endemic to Silhouette Island, at 490–1,970 ft (150–600 m) ASL. All four species are small with a cryptic patterning of browns and greens. *Sooglossus* species have more pointed heads than *Sechellophryne* species, and they lack toe pads or webbing.

DISTRIBUTION
Seychelles (Mahé, Silhouette, Praslin, and Cerf)

GENERA
Sechellophryne, Sooglossus

HABITATS
Primary rainforest and granitic ravines close to streams

SIZE
½ in (11 mm) Gardiner's Frog (*Se. gardineri*) to 2¼ in (55 mm) Thomasset's Frog (*So. thomasseti*)

ACTIVITY
Crepuscular or nocturnal; terrestrial or arboreal

REPRODUCTION
Amplexus is inguinal; eggs are laid on the ground, either

hatching into tadpoles or froglets depending on species

DIET
Small invertebrates, e.g., mites, ants, insect larvae, and amphipods

IUCN STATUS
CR = 2, EN = 2; percentage species in trouble = 100%

ABOVE | Female Seychelles Frogs (*Sooglossus sechellensis*) lay eggs that hatch into nonfeeding tadpoles which are then carried by the female until their complete metamorphosis.

OPPOSITE | Gardiner's Frog (*Sechellophryne gardineri*) is tiny—just ½ in (11 mm) as an adult—and is a direct breeder, its froglets hatching from eggs laid on the forest floor.

Reproductive strategy varies with species. Females of *So. sechellensis* and *So. thomasseti* lay small clutches of eggs on the ground, which they then protect until they hatch into nonfeeding tadpoles, which then ride on the female's back until metamorphosis is complete. *Sechellophryne gardineri* is a direct breeder; the eggs laid on the ground hatch into froglets, omitting the larval stage. The strategy of *Se. pipilodryas* is unknown, as are its dietary preferences.

Sooglossid frogs are ancient frogs, most closely related to the purple pig-nosed frogs (*Nasikabatrachus*). *Sechellophryne pipilodryas* and *So. thomasseti* are Critically Endangered, while *Se. gardineri* and *So. sechellensis* are Endangered. In 2003 the Seychelles issued a set of four postage stamps to raise awareness about their endemic frogs.

AUSTRALOPAPUAN GROUND FROGS

The Limnodynastidae and Myobatrachidae of Australasia and the Chilean Calyptocephalidae form the superfamily Myobatrachoidea.

Limnodynastidae contains 7 genera and 44 species in wide variety of habitats. The secretive, squat, bug-eyed burrowing frogs *Heleioporus* (6 spp.) inhabit coastal ranges in southern Australia, and include the largest species, the Giant Burrowing Frog (*H. australiacus*). Several species are called owl frogs because of their hooting calls, while the Moaning Frog (*H. eyrie*) makes a haunting sound. The trilling frogs *Neobatrachus* (9 spp.) issue a high-pitched call and differ from *Heleioporus* in possessing webbed toes. Three *Neobatrachus*, including the Kunapalari Frog (*N. kunapalari*), are tetraploid, possessing two pairs of each chromosome, rather than the usual diploid pair. The rotund burrowing frogs in *Notaden* (4 spp.) inhabit woodland, grassland, and desert, and includes the Crucifix Frog (*N. bennettii*), which bears a warty cross-mark. *Neobatrachus* and *Notaden* lay strings of eggs in water while other limnodynastids are terrestrial foam nest builders.

Genus *Limnodynastes* (11 spp.) occurs from Tasmania to northern Australia and southern New Guinea, where the distinctive Marbled Frog

DISTRIBUTION
Australia, New Guinea, and Aru Islands

GENERA
Adelotus, Heleioporus, Limnodynastes, Neobatrachus, Notaden, Philoria, Platyplectrum (*Lechriodus* is synonymized with *Platyplectrum*)

HABITATS
Rainforest, wet sclerophyll forest, savanna woodland, grassland, scrub, sandy or peaty swamps, and near streams or pools

SIZE
1 in (27 mm) Richmond Sphagnum Frog (*Pi. richmondensis*) to 4 in (100 mm) Western Spotted

Frog (*H. albopunctatus*) or Giant Burrowing Frog (*H. australiacus*)

ACTIVITY
Nocturnal; fossorial or terrestrial

REPRODUCTION
Amplexus is inguinal following heavy rain, strings of eggs laid in water (*Neobatrachus, Notaden*), in wet soil burrows

(*L. convexiusculus*) inhabits eucalypt woodland. Many of the Australian species are called banjo frogs because their twanging call. The genus *Platyplectrum* (6 spp.) contains three Australian species, including Fletcher's Frog (*P. fletcheri*), two New Guinea endemics, and one New Guinea-Aru Islands species, the Wokan Cannibal Frog (*P. melanopyga*), although it is the tadpoles that are cannibalistic, not the adults. The Tusked Frog (*Adelotus brevis*) is a cryptic, monotypic, eastern Australian species with a pair of long tusklike pseudo-teeth in its lower jaw; larger in males, which may use them in territorial combats. The smallest and most endangered limnodynastids are the mountain frogs, *Philoria* (7 spp.) of eastern Australia, which exhibit localized ranges in montane rainforests or wet sclerophyll forests. Inhabiting the mossy forest floor, they lay their eggs in wet soil tunnels. The Baw Baw Frog (*Philoria frosti*) of Victoria is Critically Endangered, five other *Philoria* are Endangered, *Heleioporus australiacus* is Vulnerable, and *Adelotus brevis* is Near Threatened.

OPPOSITE | The Moaning Frog (*Heleioporus eyrei*), from southwestern Australia, is named for its moaning call.

ABOVE | The Wokan Cannibal Frog (*Platyplectrum melanopyga*) is a widely distributed species in New Guinea and the Aru Islands.

BELOW | Male Tusked Frogs (*Adelotus brevis*), from eastern Australia, use the elongate pseudo-teeth in their lower jaws in territorial disputes.

(*Philoria*), or foam-nests floating on the water

DIET
Small invertebrates; some *Platyplectrum* tadpoles are cannibalistic

IUCN STATUS
CR = 1, EN = 5, VU = 1, NT = 1; percentage species in trouble = 18%

AUSTRALOPAPUAN FROGLETS & TOADLETS

The Myobatrachidae contains 13 genera (91 spp.), with its distribution centered over Australia, excluding the deserts. Ten genera (85 spp.) are continental Australian endemics, while three species are shared with southern New Guinea: the Remote Toadlet (*Crinia remota*), Smooth Toadlet (*Uperoleia lithomoda*), and Mimic Toadlet (*U. mimula*). The Namosado Barred Frog (*Mixophyes hihihorlo*) is the only New Guinea endemic, while the Moss Froglet (*Crinia nimba*) and Tasmanian Froglet (*C. tasmaniensis*) are endemic to Tasmania.

Adult *Crinia*, *Geocrinia* and *Uperoleia* are less than 1¼ in (30 mm) SVL and are known as froglets or toadlets, but the genus *Mixophyes* contains large species, for example, the Giant Barred Frog (*M. iteratus*) from coastal New South Wales and Queensland.

Myobatrachid frogs inhabit many habitats, and are terrestrial or fossorial. The Western Australian genus *Arenophryne* (2 spp.) contains rotund frogs known as sandhill frogs, after their coastal habitat. The 2½ in (60 mm) Turtle Frog (*Myobatrachus gouldii*), a fossorial species from southwestern Australia, is a bloated, short-legged, termite-eating species not dissimilar to *Rhinophrynus dorsalis* in Mesoamerica (see page 74), or *Nasikabatrachus* of southern India (see page 90). The southwestern

DISTRIBUTION
Australia and southern New Guinea

GENERA
Arenophryne, Assa, Crinia, Geocrinia, Metacrinia, Mixophyes, Myobatrachus, Paracrinia, Pseudophryne, Rheobatrachus, Spicospina, Taudactylus, Uperoleia

HABITATS
Grassland, semidesert, coastal dunes, mountain streams, rainforest, and cool wet forest

SIZE
¾ in (18 mm) Kimberley Froglet (*C. fimbriata*), Sloane's Froglet (*C. sloanei*), and Wallum Froglet (*C. tinnula*) to 4½ in (115 mm) Giant Barred Frog (*Mi. iteratus*)

ACTIVITY
Terrestrial, fossorial

REPRODUCTION
Males adopt inguinal amplexus with extremely diverse strategies and unique parental care in several species. *Rheobatrachus* incubated its eggs in its stomach

OPPOSITE | The Critically Endangered Southern Corroboree Frog (*Pseudophryne corroboree*) is so iconic it has appeared on an Australian postage stamp.

RIGHT | At 4½ in (115 mm) the Great Barred Frog (*Mixophyes iteratus*) is the largest member of its genus, which contains eight Australian and one Papuan species.

BELOW | The myrmecophagous Turtle Toad (*Myobatrachus gouldii*) is unlike any other Australian anuran but it resembles *Naskibatrachus* of India and *Rhinophyrnus* of Central America.

DIET
Small invertebrates, especially termites

IUCN STATUS
EX = 3, CR = 6, EN = 4, VU = 5, NT = 4; percentage species in trouble = 23%

peat-swamp-dwelling Sunset Frog (*Spicospina flammocaerulea*), is a striking, speckled blue-on-black species with a bright flame-orange throat, neck, lips, and limbs.

Reproductive strategies vary from free-swimming tadpoles to direct breeding on land, but it is the parental care of some species that stands out. Female barred frogs, *Mixophyes* (9 spp.), flick their eggs into rocky crevices or vegetation above the water level. Male pouched frogs, *Assa* (2 spp.), possess inguinal pouches where newly hatched tadpoles are carried until developed, while female torrent-dwelling Southern and Northern Gastric-brooding Frogs (*Rheobatrachus silus* and *R. vitellinus*) swallowed their own fertilized eggs and, turning off their gastric acids, stomach-brooded them until fully developed froglets were "birthed" through their mouths. Both *Rheobatrachus* are now Extinct, as is *Taudactylus diurnus* (Mt Glorious Torrent Frog), while four other *Taudactylus*, the black and yellow *Pseudophryne corroboree* (Southern Corroboree Frog), and *Geocrinia alba* (White-bellied Frog) are all Critically Endangered.

CHILEAN WATER TOADS & MOUNTAINS FALSE TOADS

BELOW | Bullock's Mountains False Toad (*Telmatobufo bullocki*) is a rare and Endangered inhabitant of Chile's temperate *Nothofagus* forests.

The Neotropical representative of the Myobatrachoidea is the small family Calyptocephalellidae, which contains two genera endemic to the western Andes of southern and central Chile, with a fossil record extending back to the Eocene. The Chilean Water Toad (*Calyptocephalella gayi*), from the southern Atacama south to Los Lagos, exhibits considerable sexual dimorphism, males achieving 4¾ in (120 mm) SVL while females may reach 12½ in (320 mm), the second-largest anuran in the world after the Goliath Frog (*Conraua goliath*, see page 182). This is a robust-bodied amphibian, with a rounded head, small eyes, stocky limbs, and long fingers and toes, the latter being webbed for swimming. It is a highly aquatic species that inhabits ponds and lagoons up to 3,280 ft (1,000 m) ASL, where females may lay up to 10,000 eggs. Tadpoles are slow-swimming in the relatively still waters.

DISTRIBUTION
Chile

GENERA
Calyptocephalella, Telmatobufo

HABITATS
Southern beech forest with fast-flowing montane streams (*Telmatobufo*), and ponds and lagoons (*Calyptocephalella*)

SIZE
2¾ in (70 mm) Los Queules False Toad (*Telmatobufo ignotus*) to 12½ in (320 mm) ♀ Chilean Water Toad (*Calyptocephalella gayi*)

ACTIVITY
Nocturnal (*Telmatobufo*), anytime (*Calyptocephalella*); aquatic, terrestrial, secretive

REPRODUCTION
Males adopt axillary amplexus, female *Calyptocephalella* laying 10,000 eggs in still waters, *Telmatobufo* breeding in fast streams

DIET
Presumed invertebrates

IUCN STATUS
EN = 3, VU = 1; percentage species in trouble = 80%

The other genus, *Telmatobufo* (4 spp.), contains the Chilean mountains false toads, which with their stocky, warty bodies, short limbs, and large eyes do resemble bufonid toads. The false toads are inhabitants of fast-flowing streams in montane southern beech (*Nothofagus*) forests. They have fully webbed toes for swimming and their tadpoles possess large oral suckers and powerfully muscular tails to enable them to maintain their position in the water, or swim in strong currents, even against the current. The highest elevation species is the Chilean Mountains False Toad (*T. venustus*) occurring at 4,200–5,580 ft (1,280–1,700 m) ASL, while Bullock's Mountains False Toad (*T. bullocki*) is found at 2,620–3,940 ft (800–1,200 m) ASL,

and the southernmost species, the Pelado Mountains Toad (*T. australis*), occurs at 0–3,280 ft (0–1,000 m) ASL. In 2010 a fourth species, *T. ignotus*, was described from the Las Queules National Reserve on the coast.

All *Telmatobufo* except *T. australis* are Endangered, while *Calyptocephalella gayi* is Vulnerable.

ABOVE | Known from three locations in Chile, by 2014 the Chilean Mountains False Toad (*Telmatobufo venustus*) had disappeared from two of these.

INSET | The Chilean Water Toad (*Calyptocephalella gayi*) is the second largest anuran in the world though females are almost three times the size of males.

TURKEIT HILL FROGS

The genus *Allophryne* was formerly in the Centrolenidae (glass frogs), from which it differs in lacking any teeth. Today it is allocated its own family, the Allophrynidae, which is the sister taxon to the Centrolenidae. Members of the genus inhabit northern South America and are colloquially referred to as Turkeit Hill frogs after the location below the 741 ft (226 m) Kaieteur Falls in Guyana where the holotype of Ruthven's Frog (*A. ruthveni*) was collected. There are now three species in the genus, the other two described in 2012 and 2013.

Ruthven's Frog is recorded from Venezuela, the Guianas, and Lower Amazonian Brazil. The other two species are the Resplendent Frog (*A. resplendens*), from the Upper Amazon in Brazil and Peru, and the Relict Frog (*A. relicata*), which was discovered in the fragmented northern Atlantic coastal forests of Bahia, northeast Brazil.

All three species are small, semiarboreal frogs that may be found in low vegetation or on the ground in lowland rainforest, especially that which is seasonally flooded with freshwater, where they inhabit bromeliads and other arboreal phytotelma microhabitats. They have short, pointed heads, large eyes, a body that is at its stoutest posteriorly, and long slender limbs with digits that terminate in small T-shaped discs. *Allophryne ruthveni* is dorsally brown with irregular black wavy lines, *A. relicata* is unicolor green or brown with a few black speckles, while *A. resplendens* is as would be expected, more strikingly patterned dark brown

DISTRIBUTION
Northern South America

GENUS
Allophryne

HABITATS
Seasonally flooded lowland rainforest

SIZE
¾ in (22 mm) Relict Frog (*A. relicta*) to 1 in (28 mm) Resplendent Frog (*A. resplendens*)

ACTIVITY
Semiarboreal, terrestrial

REPRODUCTION
Forming aggregations in the wet season in flooding forest, eggs laid in water

DIET
Small invertebrates

IUCN STATUS
Low Concern, no species are listed as under threat

or black with all dorsal and some ventral surfaces marked with a profusion of yellow spots.

During the wet season, as the waters rise, *A. ruthveni* is reported to form breeding aggregations in the low vegetation, with eggs deposited into the rising water. All three species are listed as Least Concern by the IUCN.

ABOVE | The type locality for tiny Ruthven's Frog (*Allophryne ruthveni*) is Turkeit Hill below the Kaieteur Falls in Guyana.

GREEN-BONED GLASS FROGS

BELOW | The Cascade Glass Frog (*Sachatamia albomaculata*) is a delicate species found from Honduras to Ecuador.

BELOW RIGHT | The Magdalena Giant Glass Frog (*Ikakogi tayrana*), from Colombia, is considered *incertae sedis* (of unknown taxonomic status) in the Centrolenidae.

The glass frogs are arboreal Latin American frogs with transparent undersides that leave their internal organs, bones, eggs, even their beating hearts, visible from outside. Dorsally glass frogs are green, often with yellow, white, or blue spots. Two subfamilies are recognized, Centroleninae and Hyalinobatrachinae, but the Colombian genus *Ikakogi* (2 spp.) is *incertae sedis*.

Although the subfamilies were defined by molecular taxonomy there are some morphological clues to identification. Centroleninae is the larger subfamily with nine genera and 115 species. They could be termed the "green-boned glass frogs" because most species possess pale green bones, except the Anomalous and Rosada Glass Frogs (*Nymphargus anomalus* and *N. rosada*), which have white bones.

DISTRIBUTION
Central and northern South America

GENERA
Centrolene, Chimeralla, Cochranella, Espadarana, Nymphargus, Rulyrana, Sachatamia, Teratohyla, Vitreorana

HABITATS
Rainforest; evergreen, deciduous, and cloud forest; in trees overhanging running water; and páramo (alpine tundra)

SIZE
¾ in (21 mm) Dwarf Glass Frog (*T. spinosa*) to 3 in (77 mm) Pacific Giant Glass Frog (*Ce. geckoidea*)

ACTIVITY
Nocturnal; arboreal

REPRODUCTION
Females deposit eggs on the upper surfaces of leaves, males guard the eggs until they hatch, the tadpoles dropping into the water

Most centrolenine glass frogs are small, ¾–1¼ in (20–30 mm) SVL, but *Centrolene* (24 spp.) contains larger species, such as the Pacific Giant Glass Frog (*C. geckoidea*). *Centrolene* inhabits the Northern Andes at elevations of 3,610–7,545 ft (1,100–2,300 m) ASL while the largest genus, *Nymphargus* (42 spp.), also occurs in the Northern Andes, up to 330 ft (1,000 m) ASL and along the Amazonian versant in Peru and Bolivia. A further five genera, containing 25 species, inhabit lower Central America and northwestern South America, to elevations of 8,200 ft (2,500 m) ASL.

Two eastern clusters of green-boned glass frogs of genus *Vitreorana* (10 spp.) occur to 5,580 ft (1,700 m) ASL in the Guiana Highlands and the Atlantic forests of southeastern Brazil. The Corupó Glass Frog (*Vitreorana uranoscopa*) occurs farthest south, in southern Brazil and northeastern Argentina.

Male green-boned glass frogs combat by dangling by their feet from the underside of a leaf as they wrestle for dominance. Female green-boned glass frogs lay their eggs on the upper surfaces of leaves where they are guarded by the males until they hatch into vermiform tadpoles which drop into the water below. Large species like *C. geckoidea* are more terrestrial, laying their eggs on streamside rocks. Nine species are Critically Endangered and 21 Endangered.

DIET
Small invertebrates

IUCN STATUS
CR = 9, EN = 21, VU = 18, NT = 7; percentage species in trouble = 46%

WHITE-BONED GLASS FROGS

In contrast to the Centroleninae most species in the Hyalinobatrachinae have white bones and could be termed "white-boned glass frogs" but again there are exceptions, such as the Sarisariñama and Taylor's glass frogs (*Hyalinobatrachium mesai* and *H. taylori*), and the genus *Celsiella*, which have green bones.

The Hyalinobatrachinae contains two genera with the nominate genus, *Hyalobatrachinium* (35 spp.), occupying most of its range, from sea level to 8,200 ft (2,500 m) ASL.

ABOVE | Sun Glass Frogs (*Hyalinobatrachium aureoguttatum*) inhabit the Pacific coastal Chocoan forests from Panama to Ecuador.

OPPOSITE | The internal organs of this Cappelle's Glass Frog (*Hyalinobatrachium cappellei*) are clearly visible through its translucent skin, the reason behind the name "glass frogs."

DISTRIBUTION
Southern Mexico and central and northern South America

GENERA
Celsiella, Hyalinobatrachium

HABITATS
Rainforest; evergreen, deciduous, and cloud forest; in trees overhanging running water; and páramo (alpine tundra)

SIZE
¾ in (19 mm) Fleischmann's Glass Frog (*H. fleischmanni*) to 1¼ in (30 mm) Plantation Glass Frog (*H. colymbiphyllum*)

ACTIVITY
Nocturnal; arboreal

REPRODUCTION
Females deposit eggs on the undersides of leaves, males

guard the eggs until they hatch, the tadpoles dropping into the water

DIET
Small invertebrates

IUCN STATUS
EN = 6, VU = 2; NT = 4; percentage species in trouble = 32%

The Northern Glass Frog (*H. viridissimum*) is the northernmost species reaching Guerrero on the Pacific coast of Mexico, and Veracruz on the Caribbean coast. *Hyalinobatrachium* is found throughout Central America and into South America as far south as Bolivia and Mato Grosso, Brazil. The second genus, *Celsiella* (2 spp.) is endemic to the northern coast of Venezuela, including the Paria Peninsula, below 5,900 ft (1,800 m) ASL.

Glass frogs of both subfamilies inhabit a variety of forested habitats from lowland rainforest to cloud forest, and even alpine tundra known as páramo, but with a preference for trees overhanging running water. Although glass frogs like rainforests, they do not like heavy rain and it has been suggested that a direct hit from a large raindrop could kill one of these small, delicate anurans.

Male white-boned glass frogs engage in a form of wrestling that resembles amplexus, on the upper surfaces of the leaves. Unlike the green-boned glass frogs, white-boned glass frog females lay their eggs on the undersides of leaves where they are guarded by the males until they hatch, a single male attending clutches laid by several females. Six species of hyalobatrachine glass frogs are considered Endangered.

TOP RIGHT | The Cerro El Humo Glass Frog (*Celsiella vozmedianoi*), from the Paria Peninsula of Venezuela, has green bones.

SOUTH AMERICAN FROGS

RIGHT | Extirpated from many of the islands where it was once common, the Mountain Chicken (*Leptodactylus fallax*) is hunted for food.

The Leptodactylidae was once a huge family with over 50 genera and 1,100 species but following the erection of 9 new families it now contains just 13 genera and around 230 species. The largest genus is *Leptodactylus* (85 spp.), the nest-building frogs. Depending on species they produce foam-nests on the water surface, or in terrestrial depressions. A female Lesser Antillean Mountain Chicken (*L. fallax*) excavates a burrow for her eggs, and guards them, laying thousands of infertile eggs to feed the developing larvae. *Leptodactylus* vary greatly in size which means several species can occur in a single location,

partitioning the resources. The large Smoky Jungle Frog (*L. pentadactylus*) even eats small vertebrates.

Adenomera (29 spp.) contains the thin-toed frogs which are terrestrial foam-nesters. Most species are small, with one of the widest-distributed being the Dark-spotted Thin-toed Frog (*A. hylaedactyla*). Genus *Hydrolaetare* (3 spp.) contains aquatic frogs, for example, Schmidt's Forest Frog (*H. schmidti*), that utilize temporary ponds and puddles across northern South America. They possess serrated fringes along their fingers and toes. The Painted Ant-nest Frog (*Lithodytes lineatus*), is a small, dark, monotypic frog with a pair of gold dorsolateral longitudinal stripes

DISTRIBUTION
Mexico, Central and South America, and Lesser Antilles

GENERA
Adenomera, Hydrolaetare, Leptodactylus, Lithodytes

HABITATS
Lowland and montane rainforest, swamps, seasonally flooded savanna, and agricultural plots

SIZE
¾ in (18 mm) Kayapo Terrestrial Foam-nest Frog (*A. kayapo*) to 7¼ in (185 mm) Smoky Jungle Frog (*Le. pentadacylus*)

ACTIVITY
Nocturnal; mostly terrestrial, or aquatic (*Hydrolaetare*)

REPRODUCTION
Amplexus axillary; most species

produce foam-nests, either in the water or in wet depressions or burrows on land

DIET
Earthworms, insects, and crustaceans; large species can take small vertebrates

IUCN STATUS
CR = 3, EN, 1, VU = 1, NT = 2; percentage species in trouble = 6%

ABOVE | The tiny leaf-litter
dwelling Dark-spotted
Thin-toed Frog (*Adenomera
hylaedactyla*) is widely distributed
across Amazonia.

LEFT | The Painted Ant-nest
Frog (*Lithodytes lineatus*) gains
protection from living in the
nests of aggressive ants and
mimicking a nurse frog with
toxic skin secretions.

from its snout to its rump, which bears a red-orange
marking. It inhabits tropical forests and builds
foam-nests in temporary pools. It obtains protection
by living and calling from the nests of aggressive
leaf-cutter ants that leave the frog unmolested.

Three leptodactylids are Critically Endangered,
including the Mountain Chicken, which is hunted
for food. Extinct on Martinique, Guadeloupe, and
St Kitts and Nevis, just 100 individuals survive on
Dominica, and only two on Monserrat. It is the
subject of a large-scale conservation project.

The other two species are the Cerro Socopo Frog
(*L. magistris*), from Venezuela, and White-lipped Frog
(*L. silvanimbus*) from Honduras, while Lutz' Thin-
toed Frog (*A. lutzi*), from Kaieteur National Park,
Guyana, is Endangered.

BRAZILIAN BROMELIAD & COASTAL FROGS

ABOVE | The Bahia Frog (*Rupirana cardosoi*) is an inhabitant of high elevation grassland streams in northeast Brazil.

The leptodactylid subfamily Paratelmatobiinae is endemic to the Atlantic coastal provinces of eastern Brazil, from Bahia south to Santa Catalina. It contains 4 genera and 14 species.

Paratelmatobius (7 spp.) contains frogs that inhabit water-logged leaf-litter and rapid flowing montane rainforest streams in the

DISTRIBUTION
Brazilian Atlantic coastal ranges

GENERA
*Crossodactylodes,
Paratelmatobius, Rupirana,
Scythrophrys*

HABITATS
High-elevation montane rainforest and grassland above the tree line, water-logged leaf-litter, temporary ponds,

or fast-flowing streams (*Paratelmatobius*), and bromeliads (*Crossodactylodes*)

SIZE
½ in (14 mm) Bokermann's Bromeliad Frog (*C. bokermanni*) to 1 ¼ in (34 mm) Bahia Frog (*R. cardosoi*)

ACTIVITY
Nocturnal; secretive

REPRODUCTION
Axillary amplexus; explosive breeders after heavy rain, females deposit large eggs on the streambed or attached to rocks (*Paratelmatobius*)

DIET
Presumed small invertebrates

IUCN STATUS
NT = 3; percentage species in trouble = 21%

Atlantic coastal mountain ranges from southern Minas Gerais to Paraná. They possess cryptic brown dorsums but bright red undersides which they display by flipping themselves upside down when threatened. The most recently described species is Segalla's Coastal Frog (*P. segallai*) from the Pico de Marumbi in Paraná state.

The genus *Crossodactylodes* (5 spp.) contains the bromeliad frogs. They are unique in that they live their entire lives in phytotelmata, water-bearing bromeliads (phyto = plant, telmata = pond) in the states from Bahia to Rio de Janeiro. A male's entire territory comprises a cluster of these arboreal plants. They are small frogs with drab dorsal and ventral patterning. Four species inhabit rainforest bromeliads but the Pico de Itambé Bromeliad frog (*C. itambe*) from Minas Gerais, inhabits high-elevation (6,023–6,765 ft [1,836–2,062 m] ASL) meadows, much farther inland than its congeners, where it is associated with bromeliads growing on rocky outcrops.

Two monotypic species are included in the Paratelmatobiinae. The Banhado Frog (*Scythrophrys sawayae*) inhabits the Serra do Mar in Paraná and Santa Catarina states where its green or brown coloration helps it blend in with the rainforest leaf-litter. The Bahia Frog (*Rupirana cardosoi*) is a small frog from the Espinhaço range in Bahia, where it inhabits high-elevation grassland streams at 3,940 ft (1,200 m) ASL. This species and two of the bromeliad frogs, Bokermann's Bromeliad Frog (*C. bokermanni*) and Izecksohn's Bromeliad Frog (*C. izecksohni*), are listed as Near Threatened.

BELOW | When threatened Segalla's Coastal Frog (*Paratelmatobius segellai*) will flip itself upside down to expose its red venter, like a European Fire-bellied Toad (*Bombina bombina*).

FOAM FROGS & DWARF SWAMP FROGS

Leiuperinae contains 5 genera and 101 species of small to medium-sized frogs. It has been recognized as a sister-taxa to the Leptodactylidae but currently it is included in that family as a subfamily. It is found from southern Mexico to southern Argentina and Chile, but Leiuperinae is primarily South American with only one species, the Túngara Frog (*Engystomops pustulatus*), extending into Central America. The Túngara Frog is a model organism used in evolutionary studies. Several species have become established on islands off northern South America and the Lesser Antilles.

Four of the genera, including the largest, *Physalaemus* (50 spp.), and smallest, *Edalorhina* (2 spp.), are foam-nest frogs; the eggs of the female and sperm of the male are combined in a secretion produced by the female and beaten into a frothy foam by the male with his hindfeet. The floating foam of the genus

LEFT | The Túngara Frog (*Engystomops pustulatus*) occurs as far north as Mexico where it is reported to have a mutualistic relationship with tarantulas of genus *Aphonopelma*.

OPPOSITE TOP | Perez' Snouted Frog *(Edalorhina perezi)* is easily recognized by its black and white undersides, orange thighs, and the tubercles above its eyes, but across its range it exhibits considerable variation in its vomerine teeth and dorsal morphology.

OPPOSITE | The Chilean Four-eyed Frog (*Pleurodema thaul*) exposes the protruding eye markings of its rump to intimidate an enemy.

DISTRIBUTION
Southern Mexico and Central and South America

GENERA
Edalorhina, Engystomops, Physalaemus, Pleurodema, Pseudopaludicola

HABITATS
Almost all habitats, from rainforest to subantarctic forest, coastal dunes to montane grassland, swamps, urban areas, etc.

SIZE
½ in (13 mm) Mustache Swamp Frog (*Ps. mystacalis*) to 2 in (50 mm) Chilean Four-eyed Frogs (*Pl. thaul*) and

Gray Four-eyed Frogs and (*Pl. bufoninum*)

ACTIVITY
Nocturnal or diurnal; aquatic, terrestrial

REPRODUCTION
Axillary amplexus; eggs laid in foam or separately in temporary puddles

DIET
Small invertebrates, insects,
and spiders

IUCN STATUS
CR = 1, EN, 1, VU = 2,
NT = 2; percentage species
in trouble = 4%

Engystomops (9 spp.) may contain antimicrobial or antiparasitic properties to protect the eggs until they hatch, the tadpoles continuing their development in the water. The dwarf swamp frogs, *Pseudopaludicola* (25 spp.), do not produce a foam-nest, the females laying their eggs directly into the water.

The genus *Pleurodema* (15 spp.) contains the four-eyed frogs, so-called because they exhibit a pair of large eye-markings on their thighs. One of the largest species is the Chilean Four-eyed Frog (*Pl. thaul*), which adopts a head-down, rear-in-the-air posture when it feels threatened. Not only do the large eye-markings intimidate a potential predator, they advertise a pair of poison glands that would make the frog a distasteful meal. The Cuyaba Dwarf Frog (*Ph. nattereri*) is also toxic and advertises this fact with a pair of huge black headlamp eye-spots on its rump. Species that display such startling eye-markings are said to be deimatic, and the Frightening Foam Froglet (*Ph. diematicus*) is named for this behavior. The El Rincon Stream Frog (*Pl. somuncurense*) is Critically Endangered.

NEOTROPICAL TORRENT FROGS

LEFT | The Agile Giant Neotropical Torrent Frog (*Phantasmarana apuana*) is a rarely encountered fast-flowing stream species that leaps into the torrent at the slightest threat.

OPPOSITE INSET | The Boraceia Tree Toad (*Hylodes phyllodes*) is a tiny frog that is endemic to Sierra do Mar in São Paulo state, Brazil.

OPPOSITE BELOW | This torrent-dwelling male Baumann's Tree Toad (*Hylodes lateristrigatus*) needs a high-pitched call to be heard over its noisy environment.

Hylodidae is confined to the Atlantic coastal forests and mountains, where its members inhabit fast-flowing montane forest streams. It comprises four genera, two of which may be termed small Neotropical torrent frogs, *Crossodactylus* (13 spp.) and *Hylodes* (26 spp.), while the monotypic Goeldi's Giant Torrent Frog (*Megaelosia goeldii*) and genus *Phantasmarana* (8 spp.) contain the giant Neotropical torrent frogs.

Crossodactylus, the spine-thumbed torrent frog, is distributed from Alagoas to Rio Grande do Sul on the Brazilian coast, with the Rio Spine-thumbed Torrent Frog (*C. dispar*) and Schmidt's Spine-thumbed Torrent Frog (*C. schmidti*) also inhabit Misiones, northern Argentina, with the latter species also in Paraguay. *Hylodes* is an endemic Brazilian genus, with Fred's Torrent Frog (*H. fredi*) endemic to the 74 sq mile [192 km²] Ilha Grande

DISTRIBUTION
Brazil, Paraguay, and Argentina

GENERA
Crossodactylus, Hylodes, Megaelosia, Phantasmarana

HABITATS
Rainforest, Cerrado, and along rocky rainforest streams

SIZE
¾ in (22 mm) ♂ Rio São Francisco Torrent Frog (*C. franciscanus*) to 5 in (122 mm) ♀ Massart's Giant Torrent Frog (*P. massarti*)

ACTIVITY
Diurnal; terrestrial, aquatic

REPRODUCTION
Axillary amplexus; eggs are laid in water with tadpoles remaining aquatic

DIET
Invertebrates, frogs, and fish (*Megaelosia, Phantasmarana*)

IUCN STATUS
NT = 1; percentage species in trouble = 2%

off Rio de Janeiro. These small cryptic frogs construct underwater nests for their eggs, which hatch into exotrophic, free-swimming tadpoles. Living in noisy aquatic habitats, some of these torrent frogs have evolved high-pitched calls.

Giant Neotropical torrent frogs may achieve over 4 in (100 mm) SVL, but they are secretive and rarely seen, leaping into the torrent at the slightest threat. The Bocaina Giant Torrent Frog (*P. bocainensis*) has not been seen since it was described in 1968, although modern eDNA techniques have detected the presence of a *Phantasmarana* in water at its type locality. Giant torrent frogs are mute, but *Phantasmarana* retain vocal sacs that are inflated as part of a visual territorial display. They inhabit forested mountains from Espiritu Santo to São Paulo. Goeldi's Giant Torrent Frog inhabits the Serra dos Órgãos in the state of Rio de Janeiro. The Agile Giant Neotropical Torrent Frog (*P. apuana*) preys on a wide range of invertebrates, smaller frogs, and fish.

Most Neotropical torrent frogs are listed as Data Deficient, with only Schmidt's Spine-thumbed Torrent Frog listed as Near Threatened, but this does not mean they are not under threat, especially the larger or more localized species.

NURSE FROGS

LEFT | From Amazonian Ecuador, the Unexpected Nurse Frog (*Alloabtes insperatus*) is threatened by logging and oil exploration.

BELOW | The brightly colored Sanguine Nurse Frog (*Allobates zaporo*) lays its eggs in the leaf-litter and when they hatch it carries the tadpoles to water on its back.

Aromobatidae was once included within Dendrobatidae (poison frogs). Containing three subfamilies, its name come from the defensive skunk-like smell produced by some species. The subfamily Allobatinae contains genus *Allobates* (60 spp.), which is distributed from Pacific coastal Colombia to Bolivia and Atlantic coastal Brazil, with the Talamanca Nurse Frog (*A. talamancae*) occurring into Central America, while the Martinique Volcano Frog (*A. chalcopis*) is endemic to Martinique.

Most *Allobates* are cryptically colored and nontoxic, but the widely distributed Brilliant-thighed Nurse Frog (*A. femoralis*) has yellowish and white stripes, black flanks, and bold orange thigh markings to advertise that it is noxious to eat. The Blue-fingered Nurse Frog (*A. caeruleodactylus*), from Amazonas, Brazil, is thought to use its vivid sky-blue

fingers to deter rival males. Most species lay eggs in leaf-litter nests, one parent transporting individual hatched tadpoles on its back to a small pool, hence "nurse frogs." However, the Martinique Volcano Frog has endotrophic, nonfeeding tadpoles that metamorphoses into froglets on land. This and three other species are Critically Endangered.

DISTRIBUTION
Central and northern South America and Martinique

GENUS
Allobates

HABITATS
Lowland to mid-montane rainforest, swamps, and marshes

SIZE
¾ in (18 mm) Chupada Nurse Frog (*A. brunneus*) to 1¼ in (33 mm) ♀ Myers' Nurse Frog (*A. myersi*)

ACTIVITY
Diurnal; terrestrial

REPRODUCTION
Amplexus cephalic or absent; small clutch of eggs laid in terrestrial nests, tadpoles

carried to water individually by parent; tadpoles of some species are nonfeeding and metamorphose on land

DIET
Forest floor invertebrates

IUCN STATUS
CR = 4, EN, 5, VU = 6, NT = 2; percentage species in trouble = 28%

LINGUAL FROGS & ROCKET FROGS

The Anomaloglossinae contains *Anomaloglossus* (32 spp.), the lingual frogs from the Guayanan region of northern South America, and *Rheobates* (2 spp.), the endemic Colombian rocket frogs. *Anomaloglossus* references a small backward-facing process on the tongue that is common to members of the genus, hence "lingual frogs." They inhabit Pantepui rainforests, an ecoregion associated with the great table-top tepuis of the Guianas, and their distribution is centered on Venezuela (16 spp., 50%).

Most species are cryptically patterned, but some are brightly colored, for example, the Endangered Golden Lingual Frog (*A. beebei*) from Kaieteur, Guyana, that breeds in rainwater collected in terrestrial bromeliads. *Rheobates*, for example, the Palmate Rocket Frog (*R. palmatus*), occur at higher elevations (< 8,200 ft [2,500 m]) in the northern Colombian Andes, inhabiting cloud forest streams.

Like *Allobates*, anomaloglossines lay terrestrial eggs and have either aquatic feeding (exotrophic) or terrestrial nonfeeding (endotrophic) tadpoles, the latter including Stephen's Lingual Frog (*A. stepheni*). Degranville's Lingual Frog (*A. degranvillei*) and Dewynter's Lingual Frog (*A. dewynteri*), from French Guiana, are Critically Endangered.

LEFT | The Endangered Golden Lingual Frog (*Anomaloglossus beebei*) is a tiny inhabitant of terrestrial bromeliads at the top of Kaieteur Falls in Guyana.

DISTRIBUTION
Northern South America

GENERA
Anomaloglossus, Rheobates

HABITATS
Montane and lowland rainforest and streams

SIZE
¾ in (17.8 mm) ♂ Vacher's Rocket Frog (*A. vacheri*) to 1 in (28 mm) Big-headed Rocket Frog (*A. megacephalus*)

ACTIVITY
Diurnal; terrestrial, semiarboreal (*A. beebei*)

REPRODUCTION
Amplexus cephalic or absent; small clutch of eggs laid in terrestrial nests, tadpoles carried to water individually by parent; tadpoles of some species metamorphose on land

DIET
Forest-floor invertebrates

IUCN STATUS
CR = 2, EN, 4, VU = 2, NT = 5; percentage species in trouble = 38%

CLOUD FROGS, SKUNK FROGS & COLLARED FROGS

The Aromobatinae contains two genera, *Aromobates* (18 spp.) and *Mannophryne* (20 spp.). *Aromobates* inhabits relatively high elevations (< 10,825 ft [3,300 m]) in the Venezuelan Mérida Andes, with two species extending into the Cordillera Oriental of Colombia. They are termed "cloud frogs" because they occur at such high elevations, for example the Venezuelan Sierra Cloud Frog (*A. serranus*). The skin of the highly aquatic Skunk Frog (*A. nocturnus*) contains pungent secretions to deter predation, but no toxins.

The genus *Mannophryne* contains the collared frogs, named for their dark throats. They principally inhabit the Mérida Andes and Paria Peninsula of Venezuela. The Trinidad Collared Frog (*M. trinitatis*) and the Tobago Collared Frog (*M. olmonae*), which is also found on Little Tobago Island, are endemic to Trinidad and Tobago.

The Venezuelan Sierra Cloud Frog and the Skunk Frog are among 14 species listed as Critically Endangered, while 12 are listed as Endangered. The Venezuelan Sierra Cloud Frog may already be extinct.

ABOVE | One of only two *Mannophryne* not listed as under threat by the IUCN, the Trinidad Collared Frog (*Mannophryne trinitatis*) is a common species in the leaf-litter of Trinidad.

DISTRIBUTION
Colombia, Venezuela, Trinidad and Tobago

GENERA
Aromobates, Mannophryne

HABITATS
Lowland and montane rainforest, cloud forest, and streams

SIZE
1¼ in (29 mm) Arp's Rocket Frog (A. *walterarpi*) to 2½ in (62 mm) Skunk Frog (A. *nocturnus*)

ACTIVITY
Diurnal and terrestrial, or nocturnal and aquatic (A. *nocturnus*)

REPRODUCTION
Amplexus cephalic or absent;

small clutch of eggs laid in terrestrial nests, tadpoles carried to water individually by parent

DIET
Forest floor invertebrates

IUCN STATUS
CR = 14, EN, 12, VU = 3, NT = 6; percentage species in trouble = 92%

POISON FROGS & ROCKET FROGS

Dendrobatidae contains frogs that sequester alkaloids—batrachotoxins and tetrodotoxins—from their invertebrate diet into their own skin and advertise their toxicity with bright aposematic patterns. Dendrobatidae contains three subfamilies.

Colostethinae comprises five genera of brightly colored poison frogs and cryptically patterned rocket frogs from Costa Rica to Bolivia, with the greatest diversity in Colombia (30 spp.). Poison frogs of *Ameerega* (29 spp.) and phantasmal poison frogs of *Epipedobates* (8 spp.) are striped, with colorful thighs, though some *Ameerega* lack stripes.

Colostethus (12 spp.), *Leucostethus* (11 spp.), and *Silverstoneia* (8 spp.) contain rocket frogs with dorsolateral stripes, and most species lack toxic alkaloids. However, tetrodotoxins are found in the skin of the Common Rocket Frog (*C. inguinalis*), and toxic alkaloids are reported from the Spot-bellied Rocket Frog (*S. punctiventris*).

The Critically Endangered Quito Rocket Frog (*C. jacobuspetersi*) occurs up to 12,500 ft (3,800 m) in Ecuador. The Marbled Poison Frog (*E. boulengeri*) and the endemic Gorgona Rocket Frog (*L. siapida*), inhabit Gorgona Island (16 miles^2 [26 km^2]) in the Pacific. Five colostethines are Critically Endangered with 11 Endangered.

ABOVE | The skin of Anthony's Poison Frog (*Epipedobates anthonyi*), from southwestern Ecuador and northwestern Peru, contains an extremely powerful nicotine-like toxin called Epibatidine.

RIGHT | From Costa Rica and Panama, the Rainforest Cricket Frog (*Silverstoneia flotator*) may constitute a complex of several species.

DISTRIBUTION
Costa Rica to northern and central South America

GENERA
Ameerega, Colostethus, Epipedobates, Leucostethus, Silverstoneia

HABITATS
Rainforest, especially along watercourses

SIZE
¾ in (16 mm) ♂ Machalilla Poison Frog (*E. machalilla*) to 2¼ in (55 mm) ♀ Three-striped Poison Frog (*A. trivittata*)

ACTIVITY
Diurnal; terrestrial

REPRODUCTION
Amplexus cephalic or absent; eggs laid in leaf-litter with parental care

DIET
Forest floor invertebrates, especially ants and beetles

IUCN STATUS
CR = 5, EN, 11, VU = 7, NT = 3; percentage species in trouble = 38%

POISON FROGS & DART-POISON FROGS

LEFT | The most poisonous frog in the world, the Golden Dart-poison Frog (*Phyllobates terribilis*), obtains its skin toxins by sequestering toxins from its forest-floor invertebrate prey. In captivity, on a diet of crickets, the frog becomes nontoxic.

RIGHT | The Dying Poison Frog (*Dendrobates tinctorius*) is found in two main color phases, an all blue *azureus* phase as here, and a black and yellow phase.

OPPOSITE RIGHT | Endemic to Peru and listed as Endangered by the IUCN, the Marañón Poison Frog (*Excidobates mysteriosus*) is brown or black with white blotches.

RIGHT | The Yellow-banded Poison Frog has the scientific name *Dendrobates leucomelas* where *leuco* means white and *melas* means black. When the first specimens arrived in Europe in the 1860s, the yellow pigment had leached out in preservative so the frog was misnamed.

Dendrobatinae contains 8 genera and 66 species of poison frogs, including the most toxic genus, *Phyllobates* (5 spp.), which are used by native Colombians to tip blowpipe darts (not arrows) for hunting. Only three Colombian *Phyllobates* species are used for tipping darts and only those qualify to be called dart-poison frogs.

The most poisonous species is the Golden Dart-poison Frog (*P. terribilis*), a large dart-poison frog which occurs in a series of morphs from yellow with black feet to lime-green. An adult frog contains sufficient batrachotoxin to kill ten adult humans, and even handling a wild specimen with bare hands could have severe consequences. The other two dangerous

DISTRIBUTION
Nicaragua to northern and central South America

GENERA
Adelphobates, Andinobates, Dendrobates, Excidobates, Minyobates, Oophaga, Phyllobates, Ranitomeya

HABITATS
Lowland and montane rainforest including tepuis, especially in riparian habitats

SIZE
¾ in (16 mm) Blue-bellied Poison Frog (*An. minutus*) to 2½ in (60 mm) ♀ Golden Dart-Poison Frog (*P. terribilis*)

ACTIVITY
Diurnal; terrestrial, arboreal

REPRODUCTION
Amplexus cephalic or absent; eggs laid in leaf-litter or aloft, adults transporting individual tadpoles to tree holes or bromeliads, female-only or joint parental care

species are the Kokeo Dart-poison Frogs (*P. aurotaenia*) and the Two-toned Dart-poison Frogs (*P. bicolor*).

The related but less toxic *Dendrobates* (5 spp.) contains the stunning Yellow-banded Poison Frog (*D. leucomelas*), from the Guayanan region, and the Green and Black Poison Frog (*D. auratus*), from Colombia to Nicaragua, and introduced to Hawaii.

DIET
Forest floor invertebrates, especially ants and beetles

IUCN STATUS
EX = 1, CR = 8, EN, 10, VU = 12, NT = 2; percentage species in trouble = 53%

Oophaga (12 spp.) contains the egg-feeding poison frogs, such as the polymorphic Strawberry Poison Frog (*O. pumilio*), which also exists in several morphs including the "blue jeans" morph, red with blue legs. These frogs are noted for their parental care. The female lays eggs on a leaf where the male keeps them moist by bringing water in his cloaca. Upon hatching the female individually carries the tadpoles to water-filled bromeliads, returning to lay unfertilized eggs as food for the tadpoles.

The Panamanian Splendid Poison Frog (*O. speciosa*) is Extinct while the monotypic Demonic Poison Frog (*Minyobates steyermarki*) and seven other species are Critically Endangered.

ROCKET FROGS, POISON FROGS & WESTERN LINGUAL FROGS

Hyloxalinae contains 3 genera and 73 species with *Hyloxalus* (63 spp.) the largest genus, containing cryptic rocket frogs with dorsal stripes. A few species are more brightly colored, such as the Endangered Sky-blue Rocket Frog (*H. azureiventris*) and the Green-striped Rocket Frog (*H. chlorocraspedus*). The skin toxins of these frogs are less toxic than those of the dendrobatines. The greatest diversity within genus *Hyloxalus* is reported from Colombia (32 spp.) and Ecuador (28 spp.), with the Choco Rocket Frog (*H. chocoensis*) extending into Panama, the Quijos Rocket Frog (*H. fuliginosus*) reaching Venezuela, and the Green-striped Rocket Frog in Acre, Brazil.

The genus *Paruwrobates* (3 spp.) contains rocket frogs from the Pacific versant of the northern Andes. They inhabit humid rainforests, but are threatened by encroaching agriculture and deforestation.

The genus *Ectopoglossus* (7 spp.) inhabits the Chocó region. They resemble the genus *Anomaloglossus* (page 115) in that they possess a backward-facing process on their tongues, so they may be called "western lingual frogs." Nine hyloxine species are Critically Endangered including two of the *Paruwrobates*.

TOP | The Sky-blue Poison Frog (*Hyloxalus azureiventris*) is named for its blue undersides.

ABOVE | A male Los Tayos Rocket Frog (*Hyaloxalus nexipus*) transporting his tadpoles to a watercourse.

DISTRIBUTION
Panama to northwestern South America

GENERA
Ectoglossus, Hyloxalus, Paruwrobates

HABITATS
Lowland and montane rainforest, especially in riparian habitats

SIZE
¾ in (19 mm) ♂ Stripe-headed Rocket Frog (*H. craspedoceps*) to 1½ in (36 mm) ♀ Drab Rocket Frog (*H. sordidatus*)

ACTIVITY
Diurnal; terrestrial, semiarboreal

REPRODUCTION
Eggs laid in leaf-litter, adults transporting individual tadpoles to tree holes or pools

DIET
Forest floor invertebrates, especially ants and beetles

IUCN STATUS
CR = 9, EN, 11, VU = 5, NT = 4; percentage species in trouble = 40%

ALSODIDAE
SPINY-CHEST FROGS, PATAGONIAN GROUND FROGS & RAPIDS FROG

Alsodidae contains three genera, the largest being *Alsodes* (19 spp.), the spiny-chest frogs. Males develop a spiny patch to maintain contact with females during amplexus. Fourteen species are Chilean endemics, the Neuquen Spiny-chest Frog (*A. neuquensis*) in an Argentine endemic, and four species inhabit both countries, between 8,200–9,840 ft (2,500–3,000 m) in temperate southern beech (*Nothofagus*) and Chilean pine (*Araucaria*) forests, and grassland. Tadpoles take two years to develop, overwintering under the ice.

Eupsophus (10 spp.) are squat Patagonian ground frogs, six Chilean endemics and four shared with Argentina, that inhabit mossy bogs. The monotypic Rapids Frog (*Limnomedusa macroglossa*) from southeast Brazil, Uruguay, and northern Argentina is marbled with greens and browns and covered in large round tubercles. Tadpoles hatch in small pools and are washed into rivers by floodwater.

The Pehuanche Spiny-chest Frog (*A. pehuenche*), Cantillana Spiny-chest Frog (*A. cantillanensis*), and Mocha Island Ground Frog (*E. insularis*) are Critically Endangered by invasive trout that eat their tadpoles. The Itchy Island Spiny-chest Frog (*A. monticola*), from the Chonos Archipelago, Chile, may be even rarer; the last person to see one alive was Charles Darwin.

TOP | Nora's Spiny-chest Frog (*Alsodes norae*) inhabits the coastal *Nothofagus* forests of Chile.

RIGHT | The Mocha Island Ground Frog (*Eupsophus insularis*) is endemic to Mocha Island, Chile, where it is threatened by habitat loss.

DISTRIBUTION
Chile, Argentina, Brazil, and Uruguay

GENERA
Alsodes, Eupsophus, Limnomedusa

HABITATS
Temperate forest, montane forest, mossy habitats, temperate grassland, and rivers and marshes

SIZE
1 ¾ in (43 mm) Contulmo Ground Frog (*E. contulmoensis*) to 3 in (75 mm) Black Spiny-chest Frog (*A. nodosus*)

ACTIVITY
Nocturnal; terrestrial, semi-fossorial, aquatic

REPRODUCTION
Inguinal amplexus with eggs laid in streams or small pools

on the forest floor; development of tadpoles is slow and may include overwintering

DIET
Insects and insect larvae

IUCN STATUS
CR = 3, EN = 8, VU = 4, NT = 1; percentage species in trouble = 53%

BRAZILIAN BUTTON & RIVER FROGS

At one time the Cycloramphidae contained Alsodidae as a subfamily, but it is now confined to two genera endemic to southeastern Brazil, from Bahia to Rio Grande do Sul.

The nominate genus, *Cycloramphus* (30 spp.), contains the Brazilian button frogs from the montane Atlantic coastal forests, from Bahia to Santa Catarina, many species exhibiting localized distributions and being therefore potentially vulnerable to habitat changes and loss. *Cycloramphus* comprises two ecomorphs. The aquatic ectomorph inhabits rocky rainforest streams and lays eggs in crevices or on wet logs, which hatch into aquatic tadpoles adapted to life on rocks in the splash zone. Included are the Granular Button Frog (*C. granulosus*) and Wandolleck's Button Frog

DISTRIBUTION
Southern Brazil

GENERA
Cycloramphus, Thoropa

HABITATS
Rainforest rocky streams or leaf-litter

SIZE
1 in (28 mm) ♂ Lutz's River Frog (*T. lutzi*) to 3 in (78 mm) ♂ Military River Frog (*T. miliaris*)

ACTIVITY
Nocturnal; terrestrial, semiterrestrial, aquatic

REPRODUCTION
Eggs laid in rocky crevices or on logs in splash zones, with semiterrestrial or semiaquatic tadpoles

DIET
Small invertebrates

IUCN STATUS
CR = 1, EN = 1, VU = 2, NT = 3; percentage species in trouble = 19%

(*C. ohausi*), with their flattened bodies and webbed hindfeet needed for aquatic life in fast-flowing water. The Bandeira Button Frog (*C. bandeirensis*) is also a member of the aquatic ectomorph but inhabits high-elevation open habitats at 8,040–9,480 ft (2,450–2,890 m). The terrestrial ectomorph is exemplified by the Blumenay Button Frog (*C. bolitoglossus*) and Alto Button Frog

(*C. eleutherodactylus*). They have a more rounded body form and short, unwebbed hindlimbs, and inhabit humid forest leaf-litter. Eggs laid on the ground hatch into non-swimming endotropic tadpoles. The Alcatrazes Button Frog (*C. faustoi*) is endemic to Ilha de Alcatrazes (0.65 miles^2 [1.7 km^2]), 22 miles (35 km) off the coast of São Paulo, and is the only Critically Endangered member of the family.

The genus *Thoropa* (7 spp.) contains the Brazilian river frogs. These have pointed heads and long legs and inhabit the splash zones along streams, laying their eggs on wet rocks. While the Brazilian River Frog (*T. saxatilis*) occurs at 985–3,280 ft (300–1,000 m) elevation, the Serra do Mar River Frog (*T. taophora*) may be salt-tolerant because it forages in the coastal intertidal zone.

ESCUERZOS, SMOOTH HORNED TOADS & BAHIA FOREST FROG

RIGHT | The Common Lesser Escuerzo (*Odontophrynus americanus*) inhabits the southern Brazilian and northern Argentine savannas, but is only seen during the wet season, being primarily fossorial.

The Odontophrynidae contains three genera, the largest being *Proceratophrys* (43 spp.), the smooth horned toads. These forest floor, leaf-mimic frogs resemble small versions of the large and voracious South American horned toads, of the genus *Ceratophrys* (Ceratophryidae, see page 150), but the smooth horned toads prey on forest floor arthropods rather than the small vertebrates. Some *Proceratophrys* have raised horns over their eyes or points on their snouts, which enhance their leaf-like appearance. They inhabit northeastern Brazil, from Ceará to Rio Grande do Sul, where they occur in rainforests and dry cerrado and caatinga habitats. Most species are Brazilian endemics but Avelino's Smooth Horned Toad (*P. avelinoi*) and Peters' Smooth Horned Toad (*P. bigibbosa*) also occur in northeastern Argentina and Paraguay. *Proceratophrys* are explosive breeders in slow-moving watercourses following heavy rains.

The genus *Odontophrynus* (10 spp.) are referred to as "escuerzos," Spanish for "toads," because they resemble bufonid true toads. *Odontophrynus* is distributed from northeast Brazil, Uruguay, and Paraguay to south-central Argentina in rainforest

DISTRIBUTION
Eastern and southern South America

GENERA
Macrogenioglottus,
Odontophrynus,
Proceratophrys

HABITATS
Rainforest ponds and streams, also gallery forest, semi-deciduous woodland, cerrado, and caatinga

SIZE
1 in (25 mm) ♂ Minute Smooth Horned Toad (*P. minuta*) to 4¼ in (110 mm) Bahia Forest Frog (*M. alipioi*).

ACTIVITY
Nocturnal; terrestrial

REPRODUCTION
Largely unknown, explosive breeder (*Proceratophrys*), amplexus initially on land, then moving to water, eggs laid on bottom of stream or pond (*Odontophrynus*)

and savanna woodlands. One of the most widely distributed species is the savanna-dwelling Common Lesser Escuerzo (*O. americanus*). This species is tetraploid: it possesses four sets of chromosomes, twice the normal diploid number. Diploid populations of *O. americanus* are treated as a separate species, the Cordoba Escuerzo (*O. cordobae*). Many escuerzos are fossorial for much of their lives, emerging to breed in shallow temporary ponds or seasonally flooded wetlands.

The largest species is the Bahia Forest Frog (*Macrogenioglottus alipioi*), a monotypic species from the Atlantic coastal forests of Brazil from Pernambuco to São Paulo. Inhabiting cocoa plantations, it has large protruding eyes that give it an almost cartoon frog appearance.

Moratoi's Smooth Horned Toad (*P. moratoi*), from Brazil, is listed as Critically Endangered.

TOP | Linhares Smooth Horned Frog (*Proceratophrys laticeps*) bears a strong resemblance to the Asian horned toads of the Megaphryninae, a case of convergent evolution.

RIGHT | Habitat loss is a threat to the Bahia Forest Frog (*Macrogenioglottus alipioi*) which has been forced to adapt to live in cacao plantations.

DIET
Small invertebrates, including insects, arachnids, myriapods, isopods, and gastropods

IUCN STATUS
CR = 1, NT = 1; percentage species in trouble = 4%

MOUTH-BROODING FROGS

The Rhinodermatidae is a small family from Chile and Argentina containing two genera and three species that inhabit Valdivian temperate southern beech (*Nothofagus*) forests, bogs, and grasslands, often close to slow-moving watercourses.

Darwin's Frog (*Rhinoderma darwinii*) was described in 1841 in honor of Charles Darwin, who collected some of the earliest specimens when he visited Chile aboard HMS *Beagle*. It has a smooth-skinned body and a triangular head that

BELOW | The male Darwin's Frog (*Rhinoderma darwinii*) ingests its tadpoles and broods them in its vocal sacs then, after six weeks, it releases fully formed froglets into the leaf-litter.

DISTRIBUTION
Chile and Argentina

GENERA
Insuetophrynus, Rhinoderma

HABITATS
Valdivian temperate forest, grassland, bogs, and near cold, slow-moving streams to 3,600 ft (1,100 m) elevation

SIZE
¾ in (22 mm) Darwin's Frog (*R. darwinii*) to 2¼ in (55 mm) Barrio's Frog (*I. acarpicus*)

ACTIVITY
Nocturnal; terrestrial, aquatic

REPRODUCTION
Females lay eggs on the forest floor, males brood the tadpoles in their vocal sacs for two to six

weeks before releasing the froglets from their mouths

DIET
Small invertebrates

IUCN STATUS
CR = 1, EN = 2; percentage species in trouble = 100%

terminates in a small but distinctive proboscis. Many Darwin's Frogs are brown in color and resemble dead leaves to escape predators, but adult males are often bright green, especially those brooding eggs in their vocal sacs.

A second species, the Northern Darwin's Frog (*R. rufum*), inhabited the coast of southern Chile, but it has not been seen since 1978, is listed as Critically Endangered, and may be extinct. Darwin's Frog is listed as Endangered, and also Critically Depleted in the IUCN's new Green List categories for species recovery.

Female Darwin's Frogs lay up to 40 eggs in the leaf-litter, where the male guards them until they begin to move. He then swallows the tadpoles and broods them in his vocal sacs for up to six weeks (female *Rheobatrachus*, see page 33, brooded their eggs in their stomachs), before allowing the froglets to hop free from his mouth. While the tadpoles develop the male is struck dumb, but he continues to feed and turns bright green. Male *R. rufum* only brood the tadpoles for two weeks before depositing them into the water.

The second genus is monotypic. Barrio's Frog (*Insuetophrynus acarpicus*) is found from sea level to 2,300 ft (700 m) ASL in the Valdivian region of southern Chile, where it is an aquatic inhabitant of streams, sheltering under stones during the day and emerging at night to feed on land. It is a more rotund and larger frog that the *Rhinoderma*. It is also listed as Endangered.

ABOVE | Endemic to the Valdivian temperate forests of Chile where it inhabits streams, Barrio's Frog (*Insuetophrynus acarpicus*) is listed as Endangered.

AMERICAN TRUE TOADS

Bufonidae contains 52 genera and 636 species, distributed almost globally, although the Cane Toad (*Rhinella marina*) was introduced to Australia and New Guinea. The "true toads" share a suite of characters—for example, male tadpoles possess a Bidder's organ on the testes that persists into adulthood; toad skulls are strongly ossified and attached to the skin; they lack teeth; and many possess bufotoxin secreting parotoid glands.

American bufonids comprise 15 genera and 368 species. The North American toads, *Anaxyrus* (25 spp.), include the familiar American Toad (*A. americanus*) and Great Plains Toad (*A. cognatus*), but also endangered species. The Wyoming Toad (*A. baxteri*) was listed as Extinct in the Wild, but now survives as two small populations following an intensive captive breeding program. It is North America's most endangered anuran.

West Indian toads, *Peltophryne* (14 spp.), inhabit the Greater Antilles, while Central American toads, *Incilius* (39 spp.), are found from southern USA, for example, the Sonoran Desert Toad (*I. alvarius*) and Gulf Coast Toad (*I. nebulifer*), south to Ecuador. The most famous *Incilius* was the Golden Toad (*I. periglenes*), from the Monte Verde Cloud Forest Reserve, Costa Rica. Males were bright orange, females variably patterned. Described in 1966 and last seen in 1989,

DISTRIBUTION
Almost worldwide, and introduced to Australasia and Oceania

GENERA
Adenomus, Altiphrynoides, Amazophrynella, Anaxyrus, Ansonia, Atelopus, Barbarophryne, Beduka, Blaira, Blythophryne, Bufo, Bufoides, Bufotes, Capensibufo, Churamiti, Dendrophryniscus, Didynamipus, Duttaphrynus, Epidalea, Frostius, Incilius, Ingerophrynus, Laurentophryne, Leptophryne, Melanophryniscus, Mertensophryne, Metaphryniscus, Nannophryne, Nectophryne, Nectophrynoides, Nimbaphrynoides, Oreophrynella, Osornophryne, Parapelophryne, Pedostibes, Pelophryne, Peltophryne, Phrynoidis, Poyntonophrynus, Pseudobufo, Rentapia, Rhaebo, Rhinella, Sabahphrynus,

RIGHT | The famous Golden Toad (*Incilius periglenes*) was one of the first notable casualties of the global amphibian crisis.

OPPOSITE LEFT | One of North America's most endangered amphibians, the Wyoming Toad (*Anaxyrus baxteri*).

OPPOSITE RIGHT | Variable Harlequin Toads (*Atelopus varius*) belong to a genus containing 99 species. Four are Extinct, 61 Critically Endangered, 14 Endangered, three Vulnerable, and two Near Threatened.

the Golden Toad has been extinct for longer than it was known to science.

The largest genus is *Atelopus* (99 spp.), the harlequin or stubfoot toads of southern Central and northern South America. They have smooth, glandless skin and bright aposematic colors advertising their skin tetrodotoxins. Some are stunning, for example, the Variable Harlequin Toad (*A. varius*) and Panamanian Harlequin Toad (*A. zeteki*).

The second-largest genus, *Rhinella* (89 spp.), includes the Cane Toad, but other *Rhinella* are

endangered. Many South American toads are tiny leaf-litter dwellers, for example, the red-bellied *Melanophryniscus* (31 spp.). *Oreophrynella* (8 spp.) inhabit tepuis, while the Brazilian coastal forest toads *Dendrophryniscus* (16 spp.) breed in phytotelma (bromeliads).

The Golden Toad and 4 *Atelopus* are Extinct, while 77 species are Critically Endangered, including 61 *Atelopus* species.

Schismaderma, Sclerophrys, Sigalegalephrynus, Strauchbufo, Truebella, Vandijkophrynus, Werneria, Wolterstorffina

HABITATS
Rainforest, woodland, grassland, desert, islands, and lakes and rivers

SIZE
½ in (12 mm) ♂ Matsés Litter Toad (*Am. matses*) to 9 in (230 mm) ♂ Cane Toad (*Rhinella marina*)

ACTIVITY
Nocturnal, diurnal; terrestrial, semi-fossorial, aquatic, arboreal

REPRODUCTION
Amplexus is usually axillary, sometimes inguinal; all strategies adopted, from aquatic larvae to direct breeding and viviparity

DIET
Invertebrates to small vertebrates

IUCN STATUS
EX = 5, EW = 2, CR = 104, EN = 75, VU = 49, NT = 23; percentage species in trouble = 41%

EURASIAN & ASIAN TRUE TOADS

Eurasian bufonids comprise 21 genera and 154 species. The familiar genus *Bufo* (21 spp.) occurs across Eurasia and includes the Common Toad (*Bufo bufo*), and the Iberian Spiny Toad (*B. spinosus*), but two former congenerics, the European Green Toad and the Balearic Green Toad, have been transferred to the genus *Bufotes* (15 spp.) as *Bufotes viridis* and *B. balearicus*. Another species moved out of *Bufo* was the monotypic *Epidalea calamita*, the Natterjack Toad. This small running toad has a fine stripe down its back and inhabits sandy heathland. The isolated British populations are afforded protection.

The largest Asian genus is *Ansonia* (37 spp.), the slender toads. They are small, long-limbed, and lack parotoid glands. They inhabit fast-flowing rainforest torrents from India to the Philippines. One of the most attractive is the Borneo Rainbow Toad (*A. latidisca*). The most frequently encountered tropical Asia toads belong to the second-largest genus, *Duttaphrynus* (27 spp.), especially the Asian Black-spined Toad (*D. melanostictus*), a native of mainland Southeast Asia that has been introduced further afield.

TOP | Highly arboreal, the Yellow-spotted Tree Toad (*Rentapia flavomaculata*) belies the usual image of an entirely terrestrial toad.

MIDDLE | Once a familiar species, the Common Toad (*Bufo bufo*) is now in decline in the United Kingdom.

LEFT | The Asian Common Toad (*Duttaphrynus melanostictus*) has been widely introduced outside its range, such as this specimen in Timor-Leste.

Many Asian bufonids are small, cryptic, leaf-litter dwellers, but the largest is the Borneo Giant River Toad (*Phrynoidis juxtasper*), reaching 4³/₄ in (122 mm) SVL. Most toads are terrestrial or aquatic, but toads from several genera are arboreal, including the Yellow-spotted Tree Toad (*Rentapia flavomaculata*) on the Malay Peninsula, and Everett's Tree Toad (*R. everetti*), Hose's Tree Toad (*R. hosii*), and the Spotted Tree Toad (*Sabahphrynus maculatus*), in Borneo. The Malabar Tree Toad (*Pedostibes tuberculosus*) of India's Western Ghats is also arboreal. The recently erected genus *Sigalegalephrynus* (5 spp.), contains the curious Sumatran arboreal and cave-dwelling toads known as "puppet toads" because of their resemblance to traditional wooden Batak *sigale-gale* puppets from northern Sumatra.

Three *Ansonia*, including the Murud Black Slender Toad (*A. vidua*), 2 *Bufo*, including the Luchun Stream Toad (*B. luchunnicus*), and 2 *Pelophryne*, including the Murud Dwarf Toad (*P. murudensis*), are among 9 Critically Endangered Eurasian bufonids, 16 species are Endangered.

ABOVE | The Kinabalu Slender Stream Toad (*Ansonia spinulifer*) inhabits low montane rainforest in Borneo.

BELOW | The Natterjack or Running Toad (*Epidalea calamita*) is totally protected in the United Kingdom and Ireland where it inhabits heathland and coastal sand dunes.

AFRICAN TRUE TOADS

African bufonids comprise 16 genera and over 110 species. The Eurasian toad genera *Bufo* and *Bufotes* are present in North Africa, but the rest of Africa is occupied by the African common toads, *Sclerophrys* (45 spp.), for example, the Mauritanian Toad (*S. mauritanica*), Kassasi's Toad (*S. kassasii*) in the Nile delta of Egypt, and the Raucous Toad (*S. capensis*) of South Africa. Stocky-bodied, short-limbed, with parotoid glands and smooth or tuberculate skin, their extreme representatives are the smooth-skinned, horned Channing's Toad (*S. channingi*) from Democratic Republic of the Congo, and the excessively warty Taï Toad (*S. taiensis*) from Côte d'Ivoire and Sierra Leone, which has huge globular parotoid glands over twice the size of its eyes. The large, common, Red Toad (*Schismaderma carens*) is another southeastern African species occurring in a variety of habitats.

OPPOSITE TOP | The Kihansi Spray Toad (*Nectophrynoides asperginis*) was endemic to one Tanzanian waterfall but is now Extinct in the Wild, surviving only as a captive-bred population.

OPPOSITE MIDDLE | The Mount Nimba Viviparous Toad (*Nimbaphrynoides occidentalis*), from West Africa, is unusual in that fertilization is internal and females give birth to toadlets.

OPPOSITE BELOW | Fenoulhet's Pygmy Toad (*Poyntonophrynus fenoulheti*) is a southern African inhabitant of rocky outcrops in woodland and savanna habitats.

BELOW | The Red Toad (*Schismaderma carens*) is a common savanna woodland species in southeast Africa.

There are numerous small toads, for example, the pygmy toads, *Poyntonophrynus* (11 spp.), and Van Dijk's toads, *Vandijkophrynus* (6 spp.) of central and southern Africa. At high elevations live the alpine toads, *Altiphrynoides* (2 spp.), in Ethiopia, and mountain toadlets, *Capensibufo* (5 spp.), in South Africa. The highest elevation record is the Bambuto Torrent Toad (*Werneria bambutensis*), from 5,740–8,530 ft (1,750–2,600 m) ASL in southwest Cameroon.

African rainforests are inhabited by forest tree toads, *Wolterstorffina* (3 spp.), in Nigeria and Cameroon, Central African tree toads, *Nectophryne* (2 spp.), and Parker's Ugandan Tree Toad (*Laurentophryne parkeri*). *Nectophryne* have toes with gecko-like lamellae that enable them to walk upside-down on smooth surfaces. They lack parotoid glands but possess skin toxins that are rapidly fatal to any other amphibians.

Reproductive strategies vary from a free-swimming aquatic tadpole stage to direct development, with eggs hatching into small toadlets. In Africa there are also live-bearing toads, for example, the Nimba Toad (*Nimbaphrynoides occidentalis*) from West Africa, and the East African viviparous toads, *Nectophrynoides* (13 spp.), which develop within the female's body after internal fertilization. The Kihansi Spray Toad (*Nectophrynoides asperginis*) from Tanzania is Extinct in the Wild. Eighteen African bufonids are Critically Endangered, and 10 Endangered.

HOLARCTIC TREEFROGS

The Hylidae comprises three subfamilies of primarily arboreal frogs with round terminal toe discs to assist them when climbing, but not all hylids are treefrogs—there are terrestrial, aquatic, and even fossorial species.

The largest subfamily is the Hylinae, containing 44 genera and over 720 species, from Canada to Argentina and across Eurasia from the Atlantic to the Pacific. The nominate genus is *Hyla* (16 spp.), seven species of which occur in Europe. The most familiar is the Common Treefrog (*H. arborea*) from Western Europe. Other European species include the Iberian Treefrog (*H. molleri*), the Stripeless Treefrog (*H. meridionalis*), also Iberian, the Italian Treefrog (*H. intermedia*), and the Tyrrhenian Treefrog (*H. sarda*) on Corsica and Sardinia. The Eastern Treefrog (*H. orientalis*) is found from Eastern Europe and Turkey to the Caucasus. All are green with only subtle differences separating them, such as the relative size of the tympanum and eye. Tropical treefrogs are nocturnal but European *Hyla* bask in the sunlight.

LEFT | Excluding the Iberian and Italian peninsulas and British Isles, the European Treefrog (*Hyla arborea*) is a common species across Western Europe.

DISTRIBUTION
North, Central, and South America, West Indies, Europe, Asia, Middle East, and North Africa

GENERA
Acris, Aplastodiscus, Atlantihyla, Boana, Bokermannohyla, Bromeliohyla, Charadrahyla, Corythomantis, Dendropsophus, Dryaderces, Dryophytes, Duellmanohyla, Ecnomiohyla, Exerodonta, Gabohyla, Hyla, Hyloscirtus, Isthmohyla, Itapotihyla, Lysapsus, Megastomatohyla, Myersiohyla, Nesorohyla, Nyctimantis, Osteocephalus, Osteopilus, Phyllodytes, Phytotriades, Plectrohyla, Pseudacris, Pseudis, Ptychohyla, Quilticohyla, Rheohyla, Sarcohyla, Scarthyla, Scinax, Smilisca, Sphaenorhynchus, Tepuihyla, Tlalocohyla, Trachycephalus, Triprion, Xenohyla

The only African hylid is the Carthaginian Treefrog (*H. carthaginiensis*), which inhabits coastal Tunisia and Algeria, while the Arabian Treefrog (*H. felixarabica*) occurs as two separate populations in the southwest Arabian Peninsula and the Levant. At least six species inhabit eastern Asia, including Hallowell's Treefrog (*H. hallowellii*) in the Ryukyu Islands.

North American *Hyla* are now in *Dryophytes* (16 spp.). Ten species inhabit the United States, with the widely distributed Gray Treefrog (*D. versicolor*) entering Canada. The Pine Barrens Treefrog (*D. andersonii*), which resembles European treefrogs, inhabits bogs and marshes on the eastern seaboard and the Gulf of Mexico. The Florida Panhandle is home to six species of *Dryophytes*, including two species that get their common names from their calls, the Bird-voiced Treefrog (*D. avivoca)* and the Barking Treefrog (*D. gratiosus*). Not all authors accept the validity of *Dryophtes* and place its species back in *Hyla*.

Mexico has seven species, including the Arboreal Treefrog (*D. arboricola*) and Walker's Treefrog (*D. walkeri*), while Guatemala has the Critically Endangered Bocourt's Treefrog (*D. bocourti*). Four species occur in Eastern Asia, including the Endangered Suweon Treefrog (*D. suweonensis*) from Korea.

HABITATS
Rainforest, dry forest, woodland, grassland, and wetlands

SIZE
¾ in (20 mm) Little Grass Frog (*Pseudacris ocularis*) to 5½ in (142 mm) ♀ Hispaniolan Giant Treefrog (*Osteopilus vastus*)

ACTIVITY
Arboreal, terrestrial, fossorial (*Smilisca, Triprion*), aquatic (*Pseudis*)

REPRODUCTION
Numerous methods; axillary amplexus; eggs laid in water, tree holes, and bromeliads, with free-swimming tadpole stage

DIET
Small invertebrates and other frogs (*Osteopilus septentrionalis*)

IUCN STATUS
CR = 43, EN = 54, VU = 39, NT = 18; percentage species in trouble = 24%

ABOVE | The Pine Barrens Treefrog (*Dryophytes andersonii*) is an American member of a genus also found in eastern Asia.

NORTH AMERICAN & WEST INDIAN TREEFROGS

The treefrog fauna of North America is dominated by the chorus frogs, *Pseudacris* (17 spp.), which occur in every contiguous US state, and extreme southern Alaska, every Canadian province except Nunavut and Labrador, and the entire length of the Baja California Peninsula, Mexico. The northernmost species is the Boreal Chorus Frog (*P. maculata*), but it is the Pacific Chorus Frog (*P. regilla*) which is established in southern Alaska. The widest-distributed species is the Spring Peeper (*P. crucifer*), found in 35 US states and 6 Canadian provinces. Chorus frogs are terrestrial or semiarboreal, perching on grasses and rushes. The Little Grass Frog (*P. ocularis*), from the southeast, is probably the smallest hyline treefrog.

Acris (3 spp.) includes Blanchard's Cricket Frog (*A. blanchardi*) in the Midwest and the Eastern Cricket Frog (*A. crepitans*) and Southern Cricket Frog (*A. gryllus*) in the southeast. The genus differs from other hyline treefrogs in possessing 22 rather than 24 chromosomes. Both genera have small toe pads and very limited webbing. The word "akris" means locust and is a reference to their calls , as is their common name, cricket frog.

LEFT | Spring Peepers (*Pseudacris crucifer*) are found in 35 US states and 6 Canadian provinces.

The West Indian treefrogs, *Osteopilus* (8 spp.), are moderate to large frogs that inhabit the Greater Antilles, such as Jamaica (4 spp.) and Hispaniola (3 spp.). Most are associated with mesic forest, but have adapted to living in plantations. Species like the Jamaican Yellow Treefrog (*O. marianae*) lay their eggs in bromeliads, while the Hispaniolan Laughing Treefrog (*O. dominicensis*) prefers standing or slow-flowing water for egg deposition and is often found on the ground.

While the Jamaican Yellow Treefrog is Endangered, its relative the Cuban Treefrog (*O. septentrionalis*), which occurs naturally in Cuba, the Cayman Islands, and the Bahamas, has been introduced to Puerto Rico, the Virgin Islands, Hawaii, and Florida where, ironically, it poses a threat to smaller frog species through predation and competition.

TOP | Blanchard's Cricket Frog (*Acris blanchardi*) occurs in the Midwest from the Great Lakes to the Gulf of Mexico.

INSET | The Cuban Treefrog (*Osteopilus septentrionalis*) is such a large species that it sometimes causes power outages when it reaches across gaps between electric cables.

PAN-NEOTROPICAL TREEFROGS

LEFT | The White-leaf Treefrog (*Dendropsophus leucophyllatus*) is a highly variable species that occurs in numerous different color patterns.

OPPOSITE | Found in woodlands and semidesert habitats from Texas to Costa Rica, the Mexican Treefrog (*Smilisca baudinii*) is a relatively large species.

BELOW | Prince Charles' Treefrog (*Hyloscirtus princecharlesi*) was described in 2007 and given its eponym to recognize the then Prince of Wales' support of rainforest conservation.

Neotropical hylid frog diversity is considerable, with almost 40 genera. Six of these occur in both Central and South America, and three of those are among the largest hylid genera: *Boana* (98 spp.), *Dendropsophus* (109 spp.), and *Scinax* (129 spp.). *Boana* contains the gladiator frogs, so-called because males bear spines on their hands to combat rivals. They occur from Nicaragua to Argentina. A few species are widely distributed, such as the Rusty Treefrog or Giant Gladiator Frog (*B. boans*), occurring from Panama to Bolivia. Others are more localized, such as the Hispaniolan Green Treefrog (*B. heilprini*), the only Caribbean species apart from three species in Trinidad.

Most hyline treefrogs have 24 chromosomes, but the dwarf treefrogs, *Dendropsophus*, found from Mexico to Argentina, possess 30 chromosomes. These are petite yellow treefrogs with their greatest diversity in Brazil (72 spp.). *Scinax* are dwarf treefrogs known as snouted treefrogs because of the broad, upturned snouts of some species. Distributed from Mexico

to Argentina, only Stauffer's Snouted Treefrog (*Sc. staufferi*) occurs north of Honduras. The most widely distributed species is the Red Snouted Treefrog (*Sc. ruber*), which occurs from Colombia to Bolivia, and on Trinidad and Saint Lucia, with introductions elsewhere.

Fringe-toed treefrogs, *Hyloscirtus* (38 spp.), are grouped together because they share several molecular similarities not seen in other treefrogs. They are found from Costa Rica to Bolivia. The fringe-limbed treefrogs, *Ecnomiohyla* (12 spp.), of Mexico to Colombia, have much more extensive

fringes along their limbs and huge hands and feet. *Smilisca* (9 spp.) inhabit arid deserts, from the US to Ecuador, where they avoid desiccation by burrowing underground. Included is the Mexican Treefrog *Sm. baudinii*), of Texas to Panama.

Twelve species are Critically Endangered, including two insular Atlantic endemics, on Ilha de Alcatrazes (*Sc. alcatraz*) and Ilha da Queimada Grande (*Sc. peixotoi*), the Mexican Upland Burrowing Treefrog (*Sm. dentata*), and the Cerro Socopo Treefrog (*D. amicorum*). It is perhaps apt that a Colombian species should be named for the musician and environmental campaigner Gordon Sumner, also known as "Sting" (*D. stingi*).

LEFT | The Sipurio Snouted Treefrog (*Scinax elaeochroa*) is a small frog that occurs from Nicaragua to Colombia.

MEXICAN & CENTRAL AMERICAN TREEFROGS

Most Mexican hylid frogs inhabit the tropical southern states, for example, Oaxaca, Guerrero, Puebla, Veracruz, and Chiapas. The largest endemic genus is *Sarcohyla* (26 spp.), meaning "fleshy treefrogs"—a reference to their thick, glandular skin. They are stream breeders associated with upland pine–oak woodlands and cloud forest at 4,920–10,170 ft (1,500–3,100 m) elevation. The large-mouthed treefrogs, *Megastomatohyla* (4 spp.), small, voiceless frogs that lack tympana, are also associated with the cloud forests and pine–oak forests of southern Mexico, as are the related ravine frogs, *Charadrahyla* (10 spp.), while the small, green, monotypic Small-eared Treefrog (*Rheohyla miotympanum*) is a widely distributed Mexican endemic occurring up to 655 ft (2,000 m) elevation.

Several Mexican genera extend into Central America, for example, the stream-breeding treefrogs *Quilticohyla* (4 spp.), bromeliad treefrogs, *Bromeliohyla* (3 spp.), highland treefrogs, *Exerodonta* (10 spp.), spike-thumbed treefrogs, *Plectrohyla* (19 spp.), and mountain stream treefrogs, *Ptychohyla* (6 spp.).

The mountain brook frogs, *Duellmanohyla* (10 spp.), named for William Duellman, an expert on Middle American hylid taxonomy, the rain treefrogs, *Tlalocohyla* (4 spp.), and the casque-headed treefrogs, *Triprion* (3 spp.), occur from southern Mexico to Guatemala or Honduras, but they are also represented in lower Central America by several species, including the Costa Rican Brook Treefrog (*D. uranochroa*), the Loquacious Treefrog (*Tl. loquax*), and the Coronated Treefrog (*Tr. spinosus*). This last species has a series of raised crests on the rear of its head that resemble a crown, while the related Mexican Shovel-nosed Treefrog (*Tr. spatulatus*) and the Yucatan Shovel-nosed Treefrog (*T. petasatus*) have long, flattened snouts so that their heads resemble snowplows. *Triprion* are xeric woodland and savanna species.

LEFT | The Coronated Treefrog (*Triprion spinosus*) has a raised bony crest across the back of its head, like a crown, which it uses to block the entrance to its burrow.

Of the Central American hylids only two genera do not occur in Mexico. The Atlantic spine-thumbed treefrogs, *Atlantihyla* (3 spp.), two Honduran and one Guatemalan endemics, and the isthmalian treefrogs, *Isthmohyla* (14 spp.), are confined to Costa Rica and Panama, except for *I. insolita* from northern Honduras. These 14 genera contain 30 Critically Endangered and 35 Endangered species.

SOUTH AMERICAN TREEFROGS

The Hylinae is represented in South America by 18 genera that do not enter Central America. The largest, *Bokermannohyla* (30 spp.), contains stream-dwelling treefrogs endemic to the Atlantic coastal ranges of Brazil. Izecksohn's Treefrog (*B. izecksohni*) is a "Lazarus species," believed extinct in 2004, but it was rediscovered two years later.

Other large genera include the slender-legged treefrogs, *Osteocephalus* (27 spp.), casque-headed canopy treefrogs, *Trachycephalus* (18 spp.), canebrake treefrogs, *Aplastodiscus* (16 spp.), and lime treefrogs, *Sphaenorhynchus* (15 spp.). The monotypic genus, *Phytotriades*, contains the Golden Treefrog (*P. auratus*) of Trinidad and the Paria Peninsula, Venezuela.

OPPOSITE TOP | The Cayenne Slender-legged Treefrog (*Osteocephalus leprieurii*) is a common northern South American treefrog.

OPPOSITE MIDDLE | The Orinoco Lime Treefrog (*Sphaenorhynchus lacteus*) is a small species with sharp lines to its head resulting in the alternative name of Hatchet-faced Treefrog.

OPPOSITE BELOW | The Paradoxical Frog (*Pseudis paradoxa*) is so called because its tadpole is very much larger than the adult frog and shrinks when it goes through metamorphosis.

LEFT | The Atlantic Coastal Forest Treefrog (*Bokermannohyla hylax*) is endemic to Brazil from Bahia to Paraná.

The harlequin frogs, *Lysapsus* (4 spp.), and swimming frogs, *Pseudis* (9 spp.), are aquatic with fully webbed feet. *Lysapsus* are small, flattened frogs with protruding eyes, that float on the surface meniscus, while *Pseudis* inhabit the littoral vegetation around ponds and marshes. The bright-green Paradoxical Frog (*P. paradoxa*) achieves an adult SVL of up to 2½ in (65 mm) but has a tadpole up to 10½ in (270 mm) in length, the paradox being that when the tadpole metamorphoses into an adult it gets smaller. The Madre de Dios frogs, *Scarthyla* (2 spp.), are also aquatic frogs from Amazonia and the Maracaibo Basin. The smaller males are semiarboreal, climbing rushes or grasses to call, the larger females remaining lower down. Small and light, they can escape danger by leaping onto the surface meniscus and skittering to safety.

Tepuis, meaning "house of the gods," are dominant features of the Guianan landscape that rise steeply from the lowlands, with flat, vegetated tops. The faunas of these isolated "sky islands" are a source of fascination, Arthur Conan Doyle basing *The Lost World* on Mount Roraima. The tepuis have their own unique frog faunas, their treefrogs belonging to the genera *Myersiohyla* (6 spp.), *Tepuihyla* (9 spp.), and *Nesorohyla kanaima*, the Canaima Tepui Treefrog.

From these genera, only Izecksohn's Treefrog is listed Critically Endangered, many South American species being listed as Data Deficient, which means their conservation status is unknown.

AUSTRALIAN TREEFROGS

BELOW | The Splendid Treefrog (*Ranoidea splendida*) is a large species from the Kimberley region of Western Australia.

Pelodryadinae contains 223 species from Australia, Tasmania, New Guinea, the Solomon Islands, Indonesia, and Timor-Leste. Authors recognize different genera and some elevate the subfamily to family status.

The largest genus, *Litoria* (103 spp.), contains 38 Australia species, like the heavy-bodied Yellow-spotted Treefrog (*L. castanea*), tiny Northern Dwarf Treefrog (*L. bicolor*), and endemic Tasmanian Treefrog (*L. burrowsae*). They are arboreal green or brown frogs, but not all *Litoria* are arboreal, or even resemble treefrogs. The terrestrial species include the Rocket Frog (*L. nasuta*), which has a pointed snout, "go-faster stripes," and powerful hindlegs, and the Javelin Frog (*L. microbelos*), which inhabits forest floor leaf-litter and resembles a juvenile ranid. Some species occupy crevices in rocky escarpment, for example, Copland's Rock Frog (*L. coplandi*), while the Kimberley Rockhole Frog (*L. aurifera*) inhabits small creeks and rain-filled rock pools in the same habitat. There is even a treefrog in the arid scrublands of Australia, the Desert Treefrog (*L. rubella*), which utilizes seasonal watercourses.

Genus *Ranoidea* (72 spp.), with 50 Australian species, contains those terrestrial species formerly placed in the genus *Cyclorana*, which some authors still recognize, and many species formerly in

DISTRIBUTION
Australasia, Melanesia, Wallacea, and Southwest Pacific, introduced to Guam, New Caledonia, and New Zealand

GENERA
Litoria, Nyctimystes, Ranoidea

HABITATS
Tropical rainforest, savanna woodland, temperate woodland, sandstone escarpments, heathland, islands, desert, cultivated habitats, and human habitations

SIZE
¾ in (16 mm) ♂ Javelin Frog (*L. microbelos*) to 5¼ in (135 mm) ♀ Giant White-lipped Treefrog (*N. infrafrenatus*)

ACTIVITY
Nocturnal, crepuscular; arboreal, terrestrial, aquatic, semi-fossorial

REPRODUCTION
Amplexus is axillary; eggs are laid in pools or streams and hatch into free-swimming tadpoles

DIET
Invertebrates

IUCN STATUS
CR = 7, EN = 6, VU = 7, NT = 3; percentage species in trouble = 10%

Litoria or *Pelodryas*, such as White's Treefrog (*R. caerulea*), a large, green, short-headed species, and the Cave-dwelling Treefrog (*R. cavernicola*) from the sandstone escarpments of the Kimberley. As with *Litoria*, *Ranoidea* contains arboreal and terrestrial species. One of the most attractive is the aquatic Green and Golden Bell Frog (*R. aurea*) of southeastern Australia. Although Vulnerable in its home range, it is introduced to New Caledonia, New Zealand, and other Pacific islands.

The Eastern and Western Water-holding Frogs (*R. platycephala* and *R. occidentalis*) are arid-adapted, stout-bodied frogs with dorsally positioned eyes and nostrils, that inhabit arid Centralian habitats, where they breed during the seasonal rains, retreating underground during arid times, their outer layers of skin forming an impenetrable barrier to water loss. They can survive for years in this state, awaiting the next rains and living off the water stored in their bladders.

MELANESIAN & INDONESIAN TREEFROGS

LEFT | The Giant White-lipped Treefrog (*Nyctimystes infrafrenata*) of New Guinea and Queensland is the largest treefrog in the world.

OPPOSITE | Only known from Kuranda, northern Queensland, the Kuranda Treefrog (*Ranoidea myolae*) is listed as Critically Endangered by the IUCN.

Many Australian species are also found in southern New Guinea, for example, White's Treefrog, Desert Treefrog, Northern Dwarf Treefrog, and Rocket Frog, but New Guinea and neighboring archipelagos are also home to a vast array of pelodryadine treefrogs.

Melanesian *Ranoidea* include the attractive and widely distributed Treasury Island Treefrog (*R. thesaurensis*), which is found from New Guinea to the Solomon Islands, and the Solomon Islands Treefrog (*R. lutea*), which occurs from Buka Island to New Georgia Island.

The genus *Litoria* is usually characterized by horizontal pupils while the third pelodryadine genus, *Nyctimystes* (44 spp.), typically has bulbous eyes with vertically elliptical pupils and lower eyelids distinctively marked with fine reticulations resembling veins.

While *Litoria* and *Ranoidea* are strongly represented in Australia, the genus *Nyctimystes* is primarily New Guinean, but one New Guinea species also found in Queensland is the Giant White-lipped Treefrog (*N. infrafrenata*). It was formerly included in *Litoria* because it has horizontal pupils and lacks palpebral reticulations, but molecular analysis places it in *Nyctimystes*. Females can achieve over 5 in (130 mm) SVL, which makes this species arguably the largest treefrog in the world. So heavy are they that during heavy rain, they leap out of trees onto the tin roofs of Papuan huts, the loud bangs sounding like tennis balls bouncing off the tin sheeting.

New Guinea treefrogs of genus *Nyctimystes* inhabit many habitats and at greatly differing elevations, from the mottled-green Papuan Big-eyed Treefrog (*N. papua*), which has been recorded at 8,530 ft (2,600 m) elevation in the Owen Stanley Range, to the gray-mottled Grant's Big-eyed Treefrog (*N. granti*), which occurs in the southern lowlands.

The Timor Treefrog (*L. everetti*) is endemic to the Lesser Sunda Islands of Indonesia and

Timor-Leste, while the Tanimbar Treefrog (*L. capitula*) is endemic to the south Moluccan islands.

Seven pelodryadine species are Critically Endangered, including the Booroolong Frog (*R. booroolongensis*) from New South Wales, and the Armored Frog (*R. lorica*) and Kuranda Treefrog (*R. myola*), both highly localized Queensland endemics.

BELOW | Aua's Treefrog (*Ranoidea auae*) is named for the daughter of a mythical Papuan chieftain called Pam.

SOUTH AMERICAN LEAF-FROGS

The South American Phyllomedusinae, treated as a family by some authors, is related to the Australasian Pelodryadinae. Phyllomedusinae contains 8 genera and 67 species from Mexico to Argentina. They are called "leaf-frogs" because most species deposit their eggs on vegetation overhanging water, a strategy reducing egg predation by fish but leaving them vulnerable to egg-eating cat-eyed snakes (*Leptodeira*). Not all phyllomedusine frogs use foliage; the Bicolor Leaf-frog (*Phrynomedusa marginata*) uses rocky crevices, while the Splendid Leaf-frog (*Cruziohyla calcarifer*) deposits eggs in water-filled tree holes.

Common leaf-frogs, *Agalychnis* (14 spp.), inhabit Central and northern South America, with the Red-eyed Leaf-frog (*A. callidryas*) a familiar species which has helped advertise dozens of products. Endemic Brazilian genera leaf-frogs include the

DISTRIBUTION
Mexico and Central and South America

GENERA
Agalychnis, Callimedusa, Cruziohyla, Hylomantis, Phasmahyla, Phrynomedusa, Phyllomedusa, Pithecopus

HABITATS
Tropical rainforest

SIZE
1½ in (37 mm) ♂ Toady Leaf-frog (*Ca. atelopoides*) to 4½ in (113 mm) ♀ Giant Waxy Monkey Frog (*Phyllomedusa bicolor*)

ACTIVITY
Nocturnal; arboreal

REPRODUCTION
Amplexus is axillary, with eggs on leaves overhanging water; some species lay in small pools or in crevices, the eggs hatching into free-swimming tadpoles

DIET
Invertebrates

IUCN STATUS
EX = 1, CR = 1, EN = 3, VU = 2, NT = 2; percentage species in trouble = 10%

coastal forest leaf-frogs, *Hylomantis* (2 spp.), shining leaf-frogs, *Phasmahyla* (8 spp.), and colored leaf-frogs, *Phrynomedusa* (6 spp.).

South American leaf-frogs in the genera *Phyllomedusa* (16 spp.) and *Pithecopus* (12 spp.) are called "monkey frogs" because they walk slowly through the vegetation, bodies raised, using their hands and feet with opposable thumbs, to grip the branches. *Pithecopus* means "apelike." Many have toxic skin secretions to avoid predation, and Amerindian hunters use the secretions of the Giant Monkey Frog (*Phyllomedusa bicolor*) to induce euphoric visions and to help them hunt. Not all skin secretions are antipredator. The Waxy Monkey Frog (*Phyllomedusa sauvagii*), sleeping through the daytime on a branch, produces a waxy secretion that it smears all over its body with its hindlimbs to prevent water loss. It also conserves moisture by excreting uric acid like a reptile.

Most *Callimedusa* (6 spp.) are arboreal, for example, the Tiger-striped Leaf-frog (*Ca. tomopterna*), but the genus also contains the Toady Leaf-frog (*Ca. atelopoides*), a drab leaf-litter species that resembles a toad.

The Spiny-knee Leaf-frog (*Phrynomedusa fimbriata*) was only known from a single locality but it is now believed Extinct, a harbinger for the potential fate of frogs with localized ranges. The Lemur Leaf-frog (*A. lemur*), from Panama and Costa Rica, is Critically Endangered, and three other phyllomedusines are Endangered.

ABOVE | Probably the most photographed frog species in the world: the iconic Common Red-eyed Leaf-frog (*Agalychnis callidryas*).

BELOW | The large Splendid Leaf-frog (*Cruziohyla calcarifer*) lays its eggs in water-filled treeholes and depressions where the tadpoles hide from aquatic predators.

HORNED FROGS & BUDGETT'S FROG

Ceratophryidae is a South American family containing three genera: *Ceratophrys* (8 spp.), *Lepidobatrachus* (3 spp.), and the monotypic *Chacophrys pierottii*, but ceratophryid frogs have a reputation for gluttony shared only by African bullfrogs (*Pyxicephalus*, see page 180), and are popular in the pet trade. They are often called "Pac-Man frogs" after the all-devouring computer game.

Most ceratophryid frogs are large, the largest being the Brazilian Horned Frog (*Ce. aurita*), from the Brazilian Atlantic coastal forests, with females achieving more than 6½ in (170 mm) SVL. Other notable *Ceratophrys* include the Amazonian Horned Frog (*Ce. cornuta*), Argentine Horned Frog (*Ce. ornata*) from the pampas of northern Argentina and southern Brazil, and Cranwell's Horned Frog (*Ce. cranwelli*) from the Chaco of Paraguay and Argentina, where its range overlaps with the Chaco Frog (*Ch. pierottii*), which extends into Bolivia and Paraguay. Two species in the genus *Lepidobatrachus*, the Paraguayan Horned Frog (*L. asper*) and Budgett's Frog (*L. laevis*), also occur in the Chaco.

DISTRIBUTION
South America

GENERA
Ceratophrys, Chacophrys, Lepidobatrachus

HABITATS
Rainforest, gallery forest, Chaco savanna woodland, arid scrubland, marshes

SIZE
2 in (50 mm) ♂ Chaco Horned Frog (*Ch. pierottii*) to 7 in (178 mm) Brazilian Horned Frog (*Ce. aurita*)

ACTIVITY
Nocturnal; terrestrial, aquatic

REPRODUCTION
Axillary amplexus; numerous eggs laid in water, tadpoles free-swimming

DIET
Vertebrates, e.g., frogs, lizards, snakes, and rodents, and large invertebrates

IUCN STATUS
VU = 1, NT = 2; percentage species in trouble = 25%

Ceratophrys are large, stocky frogs with exceedingly wide mouths and large eyes positioned on top of their heads, under a pair of fleshy horns. They are usually boldly marked with green and brown. *Chacophrys pierottii* is smaller but similar in appearance, although it lacks horns, while *Lepidobatrachus* exhibit a more flattened body shape, are usually unicolor, and lack horns because their eyes are dorsally positioned, suggestive of a predator that hunts from shallow water.

These frogs estivate underground during the dry season, emerging at the first rains to breed in pools and hunt. They are voracious, being capable of ambushing and devouring other frogs, lizards, snakes, and rodents, sometimes taking prey as large as themselves. Their mouths are armed with a row of large teeth and their jaws have a bite force comparable to a small carnivorous mammal. An extinct relative, *Beelzebufo ampinga* from Cretaceous Madagascar, may have preyed on small crocodilians and nonavian dinosaurs. The tadpoles of *Ceratophrys* and *Lepidobatrachus* are carnivorous but those of *Chacophrys* are herbivores or detritivores. One species is Vulnerable and two are Near Threatened.

PATAGONIAN WOODFROGS & ANDEAN FROGS

Batrachylidae was formerly a subfamily of the Ceratophryidae. It contains 4 genera and 12 species from southern Chile and southwestern Argentina.

Genus *Batrachyla* (5 spp.) contains the woodfrogs, which inhabit Valdivian temperate southern beech (*Nothofagus*) forests and wetlands. The Marbled Woodfrog (*B. antartandica*), Banded Woodfrog (*B. taeniata*), and Gray Woodfrog (*B. leptopus*) inhabit both countries and may occur in sympatry but maintain species integrity through the different calls of the males. The Marbled Woodfrog is distinctively patterned with green, brown, and yellow blotches. Nibaldo's Woodfrog (*B. nibaldoi*) is endemic to Chile, while the Fitzroya Frog (*B. fitzroya*), named for the native Patagonian

LEFT | Marbled Woodfrogs (*Batrachyla antartandica*) inhabit the Valdivian temperate forests of Argentina and Chile.

OPPOSITE | The Emerald Forest Frog (*Hylorina sylvatica*) is the largest batrachylid species. It changes hue from emerald green during the day to a deeper green at night.

DISTRIBUTION
Chile and Argentina

GENERA
Atelognathus, Batrachyla, Chaltenobatractus, Hylorina

HABITATS
Lowland and montane Valdivian temperate southern beech forest, wetlands, gardens, and agricultural environments

SIZE
1 in (25 mm) Marbled Woodfrog (*B. antartandica*) to 3 in (75 mm) Emerald Forest Frog (*H. sylvatica*)

ACTIVITY
Nocturnal; terrestrial, arboreal, aquatic, semiaquatic

REPRODUCTION
Amplexus axillary, sometimes inguinal; eggs are laid in water

or on land, tadpoles entering the water during flooding

DIET
Small invertebrates

IUCN STATUS
CR = 1, EN = 1, VU = 3; percentage species in trouble = 42%

cypress (*Fitzroya cupressoides*), is endemic to Isla Grande in Lake Menéndez, Argentina. Woodfrogs are found under rocks or logs, but they are also arboreal. Eggs are laid on land in areas prone to flooding, the tadpoles developing in the floodwater.

The montane genus *Atelognathus* (5 spp.) occurs at 1,640–4,290 ft (500–1,500 m) elevation on Andean slopes and in Patagonian basaltic lakes. The Patagonian Frog (*A. patagonicus*), Zapala Frog (*A. praebasalticus*), Laguna Raimunda Frog (*A. reverberii*), and Las Bayas Frog (*A. solitarius*) are Argentine endemics, while the Rio Negro Frog (*A. nitoi*) is endemic to Chile. The Patagonian Frog occurs as two morphotypes, an aquatic form with extensive webbing and vascularized skin folds to enhance aquatic respiration, and a littoral form found in drier conditions that lacks these features.

The remaining genera are monotypic and inhabit both countries. The Puerto Eden Frog (*Chaltenobatrachus grandisonae*) is a stocky, short-headed green frog with reddish tubercles, from Wellington Island, Chile, and two locations in Argentina. It lays small clusters of eggs on submerged rocks. The slender Emerald Forest Frog (*Hylorina sylvatica*) is also green but with broad red-brown stripes and long, unwebbed toes. It inhabits ponds and swamps but spends the nonbreeding season under logs.

The Patagonian Frog is Critically Endangered, while the Zapala Frog is Endangered, a major threat to their survival being the introduction of non-native trout.

ANDEAN WATER FROGS

The Telmatobiidae contains a single genus, *Telmatobius* (61 spp.), of fully or semiaquatic water frogs, which are distributed down the spine of South America, from Ecuador to Argentina and Chile. Some species, such as the Chilean Water Frog (*T. arequipensis*) and the Acancocha Water Frog (*T. jelskii*) from Peru, occur at elevations up to 14,760 ft (4,500 m), while the Amable Maria Water Frog (*T. brachydactylus*) inhabits the tributaries of Lake Junin, Peru, at 13,120–19,685 ft (4,000–6,000 m), which would make it the highest elevation amphibian in the world. Lake Junin is home to another record breaker, the Lake Junin Water Frog

(*T. macrostomus*), which with an SVL of 11¾ in (300 mm) is the largest member of the genus, and the third-largest anuran in the world.

The second largest *Telmatobius* is the Lake Titicaca Water Frog (*T. culeus*). Although it breeds in the shallows, it is a deep diver that has been recorded at a depth of 390 ft (120 m), another world record for an anuran. Its excessively baggy skin provides a large surface area for oxygen exchange, allowing it to remain submerged for prolonged periods. Lake Titicaca, straddling the Peruvian–Bolivian border, is the largest South American lake, located in an area of high human density. Its frogs are predated by

DISTRIBUTION
Andean South America

GENUS
Telmatobius

HABITATS
Andean lakes, rivers, and wetlands between 3,280 and 19,700 ft (1,000 and 6,000 m)

SIZE
2 in (48.5 mm) ♂ Yellow-bellied Water Frog (*T. ventriflavum*) to 12 in (300 mm) Lake Junin Water Frog (*T. macrostomus*)

ACTIVITY
Nocturnal; aquatic, semiaquatic

REPRODUCTION
Axillary amplexus, breeding in shallows, < 500 eggs

DIET
Aquatic invertebrates, especially beetles and crustaceans

IUCN STATUS
CR = 22, EN = 21, VU = 9, NT = 2; percentage species in trouble = 89%

OPPOSITE | The deep-diving Lake Titicaca Water Frog (*Telmatobius culeus*) is Endangered due to pollution, introduced trout, and its collection for food and the pet trade.

ABOVE | It may not actually be extinct but the Sehuancas Water Frog (*Telmatobius yuracare*) is still Critically Endangered, as are one in three members of this family.

INSET | Chismisa Water Frogs (*Telmatobius chusmisensis*) are only known from a small population in a semidesert area between 6,168 and 14,760 ft (1,880 and 4,500 m) in the Chilean Andes.

introduced rainbow trout (*Oncorhynchus mykiss*) and collected for human consumption. Other threats include pollution, habitat loss, and the fungal disease chytridiomycosis.

The Amable Maria, Lake Junin, and Lake Titicaca water frogs are all listed as Endangered, but the situation is even more serious for three Ecuadorian and seven Bolivian species that have not been seen for decades and may already be extinct, although the IUCN lists them as Critically Endangered, possibly in the hopes that they may prove to be a "Lazarus species," like the Sehuencas Water Frog (*T. yuracare*) of Bolivia. Extinct in the Wild, it was known from only a single captive male, but in 2019 the species was rediscovered in nature.

NEOTROPICAL MARSUPIAL TREEFROGS & HORNED TREEFROGS

Hemiphractidae contains 6 genera and 119 species that are united by their mode of reproduction. They are called "backpack frogs" or "marsupial frogs" because the females carry the fertilized eggs on their backs, albeit in two different ways.

Females of the genera *Flectonotus* (2 spp.) from Colombia, Venezuela, and Trinidad, *Fritziana* (7 spp.) from Atlantic coastal Brazil, and *Gastrotheca* (77 spp.), which are widely distributed from Costa Rica to Argentina, are called marsupial frogs because the females possess specialized pouches on their backs in which their eggs develop. Some species hatch as well-developed tadpoles while others are direct breeders producing small froglets.

LEFT | The female Puerto Cabella Marsupial Frog (*Flectonotus pygmaeus*) deposits advanced tadpoles into a water-filled bromeliad.

DISTRIBUTION
Southern Central America and Andean South America

GENERA
Cryptobatrachus, Flectonotus, Fritziana, Gastrotheca, Hemiphractus, Stefania

HABITATS
Lowland and montane rainforest

SIZE
¾ in (19 mm) ♂ Fitzgerald's Marsupial Frog (*Fl. fitzgeraldi*) to 3¼ in (81 mm) ♀ Spix's Horned Treefrog (*H. scutatus*)

ACTIVITY
Nocturnal; semiarboreal, arboreal, terrestrial

REPRODUCTION
Females carry eggs on their back that are released as

free-swimming tadpoles or froglets

DIET
Rainforest arthropods, mollusks, and earthworms, or other frogs and lizards (*Hemiphractus*)

IUCN STATUS
CR = 6, EN = 24, VU = 19, NT = 9; percentage species in trouble = 49%

The other three genera, *Cryptobatrachus* (5 spp.) from Colombia, *Hemiphractus* (9 spp.) from the Andes, and *Stefania* (19 spp.) from the Guianas, are the backpack frogs. Females lack pouches and carry their eggs in hollows on their backs. They are direct breeders, producing froglets that remain on the female's back until more developed.

Hemiphractid frogs are nocturnal inhabitants of lowland and montane rainforest. The terrestrial or semiarboreal genus *Hemiphractus* contains the bizarre horned treefrogs that possess large, triangular, casque-like heads that begin with a fleshy nasal protuberance and terminate as a pair of large backward-facing crests, giving species like the Banded Horned Frog (*H. fasciatus*) the appearance of a small dragon. They also possess large tympana, large forward-facing eyes surmounted by fleshy tubercles, and an upper jaw of fang-like teeth— all exacerbating the effect. These armored anurans prey on other frogs and small lizards.

Gastrotheca contains a diverse array of species, including Günther's Marsupial Frog (*G. guentheri*), which is unique in that it is the only living frog species known to possess true teeth in both jaws. Frogs are believed to have lost the teeth of the lower jaw 200 MYA, and Dollo's Law (of Irreversibility) states that once lost, a structure cannot re-evolve. Günther's Marsupial Frog breaks the rule: its lower jaw is full of tiny fang-like teeth. Four *Gastrotheca* and two *Hemiphractus* are Critically Endangered.

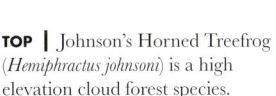

TOP | Johnson's Horned Treefrog (*Hemiphractus johnsoni*) is a high elevation cloud forest species.

MIDDLE | Endangered Horned Marsupial Frogs (*Gastrotheca cornuta*) carrry their eggs in pouches on their backs.

RIGHT | The Backpack Frog (*Stefania evansi*) carries her eggs in little hollows.

BRAZILIAN SADDLEBACK & PUMPKIN TOADLETS, FLEA FROGS & ROBBER FROGS

LEFT | The tiny Izecksohn's Pumpkin Toadlet (*Brachycephalus izecksohni*) stands out in the leaf-litter but predators would be well advised to avoid it because of its lethal skin toxins.

OPPOSITE | Holt's Robber Frog (*Ischnocnema holti*) is a common and widely distributed leaf-litter frog in the montane forests of Rio de Janeiro state, Brazil.

Brachycephalidae is the nominate family in the South American superfamily Brachycephaloidea, also known as Terrananae. It contains 2 genera and 77 species endemic to the Atlantic coastal forests and *Araucaria* forests of southeastern Brazil.

Genus *Brachycephalus* (38 spp.) contains the Brazilian saddleback toadlets, also called pumpkin toadlets, or flea frogs, the last because they are small. There are two morphotypes, the smooth-skinned flea frogs that are cryptically patterned to escape detection, and the rugose-skinned pumpkin toadlets that are brightly colored to advertise the tetrodotoxin-like neurotoxins in their skin secretions. Most species demonstrate a reduction in finger and toe number.

Pumpkin toadlets and flea frogs live in forest floor leaf-litter but occasionally they climb into low vegetation. Females lay a few large eggs in the leaf-litter and development is direct; the eggs hatch into small toadlets.

DISTRIBUTION
Southeastern Brazil and northeastern Argentina

GENERA
Brachycephalus, Ischnocnema

HABITATS
Lowland and montane rainforest

SIZE
¼ in (8.4 mm) Flea Frog (*B. pulex*) to 2 in (54 mm)

Günther's Robber Frog (*I. guentheri*)

ACTIVITY
Diurnal, possibly also nocturnal; terrestrial, semi-fossorial, arboreal

REPRODUCTION
Inguinal, changing to axillary amplexus; eggs laid in leaf-litter hatching to froglets by direct development

DIET
Small forest-floor invertebrates, e.g., insects, spiders, myriapods, crustaceans, and mollusks

IUCN STATUS
NT = 3; percentage species in trouble = 4%

The flea frogs include the Two-toed Flea Frog (*B. didactylus*), Yellow-spotted Flea Frog (*B. sulfuratus*), and the Flea Frog (*B. pulex*, *pulex* = flea), which has the smallest known maximum adult size (¼ in [8.4 mm] SVL) in the family. Among the aposematic pumpkin frogs are Alípio's Pumpkin Toadlet (*B. alipioi*), Spix's Pumpkin Toadlet (*B. ephippium*), and Izecksohn's Pumpkin Toadlet (*B. izecksohni*), all of which resemble tiny versions of the extinct Golden Toad (*Incilius periglenes*, see pages 128–9) of Costa Rica.

The genus *Ischnocnema* (39 spp.) contains the Brazilian robber frogs, which would also be endemic to southeastern Brazil but for Hensel's Robber Frog (*I. henselii*), which also occurs in northern Misiones, Argentina. The Brazilian robber frogs are cryptically patterned for life in the leaf-litter, but they may also be arboreal.

Günther's Robber Frog (*I. guentheri*) and Girard's Small Robber Frog (*I. parva*) are direct developers like the saddleback toadlets and they also demonstrate similar reductions in digit number to *Brachycephalus*. No species in these two genera is currently listed in a more serious category than Near Threatened, but many species are Data Deficient, which may obscure their true conservation status.

EMERALD-BARRED FROGS

Ceuthomantidae, containing just genus *Ceuthomantis* (4 spp.), is one of the smallest neobatrachian families. They are associated with the flat-topped tepuis of the ancient Guiana Shield in Venezuela, Guyana, and northern Brazil.

The Cerro Aracamuni Emerald-barred Frog (*C. aracamuni*) inhabits 4,898 ft (1,600 m) Cerro Aracamuni, in the Sierra de la Neblina; 75 miles (120 km) east the Pico Tamacuari Emerald-barred Frog (*C. cavernibardus*) occurs in montane forests at 3,050–4,165 ft (930–1,270 m) on Pico Tamacuari, in the Sierra Tapirapecó; 217 miles (350 km) north the Sarisariñama Tepui Emerald-barred Frog (*C. duellmani*) occurs at 3,610–5,410 ft (1,100–1,375 m), and 280 miles (450 km) east, on the Guyana–Brazil border, the Wokomung Emerald-barred Frog (*C. smaragdinus*) inhabits Mount Ayanganna and Mount Kopinang.

Emerald-barred frogs are green in coloration, if only as streaks; their digital discs are notched; they have narrow heads; they lack vomerine teeth, and males lack nuptial pads for grasping females during amplexus. They are also unusual in that they call during the day, although the Pico Tamacuari Emerald-barred Frog is reported to sing from caves. Two species are Vulnerable, one Near Threatened, and one Data Deficient.

LEFT | Wokomung Emerald-barred Frogs (*Ceuthomantis smaragdinus*) inhabit just two mountains on the border between Guyana and Brazil.

DISTRIBUTION
Venezuela, Brazil, and Guyana

GENUS
Ceuthomantis

HABITATS
Wet montane rainforest, tepuis, and caves

SIZE
¾ in (20 mm) Wokomung Massif Emerald-barred Frog (*C. smaragdinus*) to 1¼ in (32 mm) Pico Tamacuari Emerald-barred Frog (*C. cavernibardus*)

ACTIVITY
Diurnal; terrestrial, cavernicolous

REPRODUCTION
Large eggs are laid on land, probably hatching directly into froglets without a tadpole stage, but not observed

DIET
Not known, presumed invertebrates

IUCN STATUS
VU = 2, NT = 1; percentage species in trouble = 75%

ROBBER FROGS, RAIN FROGS & DIRT FROGS

RIGHT | Evergreen Robber Frogs (*Craugaster gollmeri*), from Costa Rica and Panama, adapt well to disturbed habitats, but other members of the genus are less adaptable and are Extinct, Critically Endangered, or Endangered.

The Craugastoridae contains two genera, although some authors synonymize Strabomantidae within Craugastoridae. Craugastorids are referred to as robber frogs, rain frogs, dirt frogs, or stream frogs. The nominate genus, *Craugastor* (126 spp.), inhabits Mexico and Central America, with the Barking Frog (*C. augusti*) also occurring in southwestern USA. The southernmost species is the Long-snouted Robber Frog (*C. longirostris*) of Pacific coastal Ecuador. They occur from sea level, for example, the Yucatan Robber Frog (*C. yucatanensis*), to 10,630 ft (3,240 m) elevation for the Pine-oak Robber Frog (*C. saltator*) in Guerrero, Mexico.

Genus *Haddadus* (3 spp.) is endemic to Brazil's Atlantic coastal forests with the Iguarasse Robber Frog (*H. plicifer*) in Pernambuco, and the Clay Robber Frog (*H. binotatus*) found from Bahia to Rio Grande do Sul.

Most craugastorids are direct breeders, but the Broad-headed Rainfrog (*C. laticeps*) may miss the egg stage and birth live tadpoles. The Yucatan Robber Frog is also arboreal.

The Corquin Robber frog (*C. anciano*), from Honduras, is Extinct, 15 *Craugastor* species are Critically Endangered, and 10 Endangered.

DISTRIBUTION
Southern USA, Mexico, Central America, Colombia, Ecuador, and Brazil

GENERA
Craugastor, Haddadus

HABITATS
Lowland and montane rainforest, streams, and rocky outcrops

SIZE
½ in (13 mm) Candelaria Loxicha Robber Frog (*C. candelariensis*) to 2¾ in (70 mm) Big-headed Robber Frog (*C. megacephalus*)

ACTIVITY
Nocturnal, diurnal; terrestrial, with some species also arboreal (*C. andi, C. batrachylus, C. yucatanensis*)

REPRODUCTION
Direct breeders laying eggs in leaf-litter that hatch into froglets; some birth tadpoles without an egg stage (*C. laticeps*)

DIET
Leaf-litter invertebrates

IUCN STATUS
EX = 1, CR = 15, EN = 10, VU =7, NT = 1; percentage species in trouble = 25%

NEOTROPICAL RAINFROGS, ROBBER FROGS & ANDES FROGS

Some authors place the strabomantid genera within Craugastoridae, while others recognize Strabomantidae as a family with four subfamilies.

Subfamily Holoadeninae is distributed widely across South America, at moderate to high elevations (< 13,780 ft [4,200 m]) in the Andes. Genera include *Microkayla* (25 spp.), from Bolivia and Peru, the endemic Peruvian *Bryophryne* (11 spp.), *Psychrophrynella* (5 spp.), and *Qosqophryne* (3 spp.). All are small, cryptic, direct-breeding, leaf-litter frogs that inhabit cloud forest, elfin forest, and puna grasslands. *Bryophryne* means "moss toad," while *Psychrophrynella* means "small frost toad," references to their High Andes homes. *Qosqophryne* are "Cusco toads," from "Qosqo" in Quechua, while *Microkayla* is "small frog," using the Quechua "kayla" for frog. These are small, cold-adapted frogs and most are cryptically colored but some individuals in a population may be bright yellow, such as Illiman's Andes Frog (*Microkayla illimani*).

Noblella (17 spp.) also inhabits the Andes, from Ecuador to Bolivia, but the tiny Loreto Leaf Frog (*No. myrmecoides*) occurs in the Amazonian lowlands. Male Pygmy Andes Leaf-litter Frogs (*No. pygmaea*) are the smallest Andean frogs.

BELOW TOP | Illiman's Andes Frog (*Microkayla illimani*) is a tiny cold-adapted species from a single Bolivian valley. Usually cryptically patterned, some specimens are bright yellow.

Key: (1) Holoadeninae (red) (2) Hypodactylinae (blue)

DISTRIBUTION
Northern South America

GENERA
(1) *Bahius, Barycholos, Bryophryne, Euparkerella, Holoaden, Microkayla, Noblella, Psychrophrynella, Qosqophryne;* (2) *Niceforonia*

HABITATS
Lowland rainforest, secondary forest, gallery forest, wet montane forest, elfin forest, Cerrado, and cocoa plantations (*Bahius*); cloud forest and high-elevation grassland (*Niceforonia*)

SIZE
(1) ½ in (11 mm) ♂ Pygmy Andes Leaf-litter Frog (*No. pygmaea*) to 1¾ in (48 mm) ♀ Hole-dwelling Granular Frog (*H. pholeter*); (2) ¾ in (21 mm) Santander Andes Frog (*Ni. nana*) to 2½ in (58 mm) ♀ Putumayo Robber Frog (*Ni. dolops*)

ACTIVITY
Diurnal, nocturnal; terrestrial in leaf-litter and bromeliads

REPRODUCTION
Females lay and guard small clutches of large eggs so believed to be direct breeders

DIET
Leaf-litter invertebrates

IUCN STATUS
Holoadeninae CR = 15, EN = 7, VU = 11, NT = 1; percentage species in trouble = 38%.
Hypodactylinae EN = 5, VU = 3, NT = 1; percentage species in trouble = 60%

Three genera inhabit southeast Brazilian Atlantic coastal forests, *Euparkerella* (5 spp.), the Guanabara frogs, and *Holoaden* (4 spp.), the glandular frogs. *Euparkerella* are small stout frogs with smooth skin and a pointed snout, while *Holoaden* are larger with blister-like dorsal glands. The tiny, cryptically patterned, monotypic Bahia Robber Frog (*Bahius bilineatus*) inhabits bromeliads.

Genus *Barycholos* (2 spp.) contains the Beautiful Chimbo Frog (*B. pulcher*) from lowland Pacific Ecuador and Ternetz's Chimbo Frog (*B. ternetzi*) from scattered locations in south-central Brazil. Both are small, cryptic, leaf-litter species. Within Holoadeninae ten species are Critically Endangered—eight *Microkayla* and one each from *Holoaden* and *Psychrophrynella*.

Subfamily Hypodactylinae contains *Niceforonia* (15 spp.), named for the French Catholic priest Brother Nicéforo María (1888–1980), who devoted much of his life in Colombia to the study of reptiles and amphibians. These short-headed, stout-bodied, cryptically patterned frogs are found from the Upper Amazonian foothills to the High Andes (3,280–12,630 ft [1,000–3,850 m]). Five species are Endangered.

NEOTROPICAL RAINFROGS, ROBBER FROGS & ANDES FROGS

BELOW | The Special Litter Frog (*Pristimantis espedeus*) is a tiny rainforest species from French Guiana and Suriname on the ancient Guiana Shield.

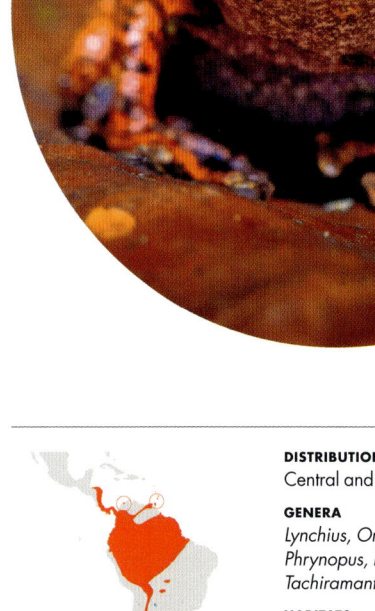

Subfamily Pristimantinae contains seven genera, most of them small in size, but the nominate genus *Pristimantis* is the largest vertebrate genus in the world with 603 species, and new species being described at an average of 19 per year. Generically referred to as rainfrogs or robber frogs, *Pristimantis* are found from Honduras to northern Argentina. They are extremely variable frogs, some resembling treefrogs and others ranids, while some are toad-like. Two species are endemic to the island of Tobago, the Charlotteville Litter Frog (*Pr. charlottevillensis*) and the Tobago Litter Frog (*Pr. turpinorum*), while Urich's Litter Frog (*Pr. urichi*) also occurs on Trinidad. The Grenada Litter Frog (*Pr. eupronides*) and the Saint Vincent Frog (*Pr. shrevei*) are endemic to the Lesser Antilles.

Often, several *Pristimantis* species occur in sympatry but occupy different niches. Most species are terrestrial despite their dilated toe and finger-tips, but the genus also contains arboreal species, such as the Salidero Rainfrog (*Pr. subsigillatus*) in Colombia and Ecuador, and Padial's Green Rainfrog (*Pr. padiali*) from Loreto, Peru.

DISTRIBUTION
Central and South America

GENERA
Lynchius, Oreobates, Phrynopus, Pristimantis, Tachiramantis, Yunganastes

HABITATS
Lowland and montane rainforest, deciduous forest, cloud forest, páramo, cerrado grassland, and Pantanal wetland

SIZE
½ in (14.5 mm) ♂ Oxapampa Andes Frog (*Ph. auriculatus*) to 2¾ in (72 mm) Zurucuchu Rainfrog (*Pr. w-nigrum*)

ACTIVITY
Diurnal, nocturnal; terrestrial, arboreal

REPRODUCTION
Direct-breeding frogs that lay small clutches of large eggs that hatch into froglets

DIET
Forest floor invertebrates

IUCN STATUS
R = 41, EN = 125, VU = 81, NT = 43; percentage species in trouble = 42.5%

The second-largest genus is *Phrynopus* (35 spp.) containing the endemic Peruvian Andes frogs which occur at high elevations, the highest record being 14,730 ft (4,490 m) for Chaparro's Andes Frog (*Ph. chaparroi*). Some species are squat with rugose skin resembling toads; others have smooth skin. The mountain frogs of the genus *Oreobates* (26 spp.) inhabit the Andes from Colombia to Argentina, while the related genus *Lynchius* (8 spp.) is confined to the Andes of Peru and Ecuador. *Tachiramantis* (7 spp.) contains the Táchira treefrogs from the river valley of the same name on the Colombia–Venezuela border, while *Serranobatrachus* (7 spp.) is endemic to the isolated Sierra Nevada de Santa Marta range on the Caribbean coast of Colombia.

Although 41 species are Critically Endangered, 125 Endangered, 81 Vulnerable, and 43 Near Threatened in the Pristimantinae, the overall percentage of species under threat, according to the IUCN, is only 42.5%, a far smaller percentage than in smaller strabomantid subfamilies.

TOP | Common Big-headed Frogs (*Oreobates quixensis*) are a widely distributed Upper Amazonian species.

LEFT | The Tayrona Frog (*Tachiramantis tayrona*) may have been the model for the gold frogs found in pre-Colombian archaeological sites.

ROBBER FROGS & ANDES FROGS

LEFT | A common species on Gorgona Island off the Pacific coast of Colombia, the Rusty Robber Frog (*Strabomantis bufoniformis*) is very rare on the mainland.

Strabomantinae contains just the genus *Strabomantis* (15 spp.) from northern South America. The Puerto Cabello Robber Frog (*S. biporcatus*) is endemic to coastal Venezuela, while the largest species, the Rusty Robber Frog (*S. bufoniformis*), is found as far north as southern Costa Rica and on Colombia's Gorgona Island.

Strabomantis are stout-bodied with short heads, short limbs, and rugose dorsal skin, characteristics reflected in the species name *bufoniformis*, which means "toad-like." They inhabit montane regions below 8,200 ft (2,500 m) elevation, although the Rusty Robber Frog is a lowland species from 50–985 ft (15–300 m), and the Dabubio Robber Frog (*S. zygodactylus*), from Pacific lowland Colombia, occurs at 750–2,620 ft (230–800 m).

Although believed direct breeders, they are little-studied, and the terrestrial nest of the Choco Robber Frog (*S. anomalus*) was documented in an area likely to be flooded, an unusual location for a direct-breeding frog.

The Nutibara Robber Frog (*S. cadenai*) from Colombia, and the Warty Groundfrog (*S. helonotus*) from Ecuador, are Critically Endangered, while the Rusty Robber Frog, Rio Calles Robber Frog (*S. cheiroplethus*), and Ruiz's Robber Frog (*S. ruizi*) are Endangered.

DISTRIBUTION
Central and South America

GENUS
Strabomantis

HABITATS
Lowland, submontane and montane rainforest, cloud forest, rocky streams, and swamps

SIZE
1¼ in (33 mm) ♂ Anatipes Robber Frog (*S. anatipes*) to 4 in (106 mm) ♀ Rio Calles Robber Frog (*S. cheiroplethus*)

ACTIVITY
Nocturnal; terrestrial

REPRODUCTION
Presumed to be direct breeders but some females lay eggs in areas that flood

DIET
Forest floor invertebrates

IUCN STATUS
CR = 2, EN = 3, VU = 4; percentage species in trouble = 56%

SHIELD FROGS & FLEA FROGS

Eleutherodactylidae contains two subfamilies, the smallest, Phyzelaphryninae, containing just two genera, *Adelophryne* (12 spp.) and *Phyzelaphryne* (2 spp.), translating as "hidden toads" and "shy toads," both references to the secretive lifestyles of these small, cryptic, leaf-litter frogs.

The Guiana Shield Frog (*A. gutturosa*) occurs across this ancient region from Venezuela to French Guiana, the Patamona Shield Frog is endemic to Guyana, and the Amapá Shield Frog (*A. amapaensis*) occurs in Amapá state, Brazil, also part of the Guiana Shield. Those *Adelophryne* occurring outside this region are called "flea frogs" because of their small size. They include the Yapima Flea Frog (*A. adiastola*) from the Upper Amazon and eight Brazilian Atlantic coastal forest species.

Miriam's Frog (*Phyzelaphryne miriamae*) and the Nimio Frog (*P. nimio*) are plentiful where they occur, but also very small. Miriam's Frog was described in 1977 and is recorded from the basins of the Upper Amazon and associated rivers. The Nimio Frog was described in 2018, from the Rio Japurá, also in Amazonia.

These small leaf-litter frogs are direct breeders, females laying a small number of large eggs on land that hatch into tiny froglets. The Serra de Maranguape Flea Frog (*A. maranguapensis*) is Endangered.

LEFT | The Sharp-fingered Flea Frog (*Adelophryne mucronata*) is endemic to the Brazilian state of Bahia.

DISTRIBUTION
Northern South America

GENERA
Adelophryne, Phyzelaphryne

HABITATS
Lowland and low montane rainforest

SIZE
½ in (9 mm) ♂ Michelin Shield Frog (*A. michelin*) to 1 in (23 mm) Patamona Shield Frog (*A. patamona*)

ACTIVITY
Nocturnal; terrestrial

REPRODUCTION
Direct breeders, females laying a few large eggs that hatch directly into small froglets

DIET
Invertebrates including beetles and ants

IUCN STATUS
EN = 1, VU =2; percentage species in trouble = 14%

NEOTROPICAL RAINFROGS, DINK FROGS, CHIRPING FROGS & COQUIS

Subfamily Eleutherodactylinae contains two genera, *Diasporus* (17 spp.) and *Eleutherodactylus* (206 spp.), although the latter genus was once very much larger. *Diasporus* contains the dink frogs, named for the male's call. They are small, direct-breeding leaf-litter dwellers with distributions centered on Panama, but their range extending to Honduras, with the Common Dink Frog (*D. diastema*), and Pacific coastal Ecuador, with the Esmeraldas Robber Frog (*D. gularis*).

ABOVE | Native to Cuba, the Greenhouse Frog (*Eleutherodactylus planirostris*) is an invasive species which has spread through the Caribbean and as far as Guam and China.

LEFT | The Esmeraldas Robber Frog (*Diasporus gularis*) is a common species in pristine and disturbed habitats.

DISTRIBUTION
Southern USA, Mexico, Central America, Northwest South America, and West Indies, introduced to Hawaii and elsewhere

GENERA
Diasporus, Eleutherodactylus

HABITATS
Lowland and montane rainforest, temperate woodland, wetlands, caves, and islands

SIZE
¼ in (8.5 mm) Yellow-striped Dwarf Frog (*E.* [*Euhyas*] *limbatus*) or Monte Iberia Dwarf Frog (*E.* [*E.*] *iberia*) to 3½ in (88 mm)

♀ Diquini Robber Frog (*E.* [*Pelorius*] *inoptatus*)

ACTIVITY
Nocturnal; terrestrial, arboreal, aquatic

REPRODUCTION
Direct breeding, most laying eggs hatching into froglets, some viviparous (*E.* [*E.*] *jasperi*)

Genus *Eleutherodactylus*, is divided into five subgenera of rainfrogs and chirping frogs. The nominate subgenus *Eleutherodactylus* (57 spp.) occurs from the Bahamas to the Lesser Antilles and contains one of the most notorious frog species, the Coqui (*E. [Eleutherodactylus] coqui*, see page 23), a native of Puerto Rico that has been introduced widely across the Caribbean and to Costa Rica, Florida, and Hawaii. Although of small size, introduced Coquis can exist in huge numbers (over 10,000 per acre) and consume large numbers of invertebrates (114,000 a night), so they may destabilize fragile island food webs. The Lesser Antillean Frog (*E. (El.) johnstonei*) has been introduced to South America as far south as São Paulo, Brazil. The subgenus *Euhyas* (98 spp.) occurs across the Greater Antilles with two outliers, the Bahamian Flat-headed Frog (*E. (Euhyas) rogersi*), and the Virgin Islands Yellow Frog (*E. (Eu.) lentus*).

Subgenus *Pelorius* (9 spp.) is endemic to Hispaniola (Haiti and the Dominican Republic) and includes some of the largest *Eleutherodactylus*, such as the Diquini Robber Frog (*E. (Pelorius) inoptatus*). *Pelorius* inhabit upland woodland, caves (Spiny Giant Frog, *E. (P.) nortoni*) or underground burrows (Eastern Burrowing Frog, *E. (P.) ruthae*). The sole member of the subgenus *Schwartzius*, the Yellow Cave Frog (*E. (Sc.) counouspeus*), is endemic to the limestone caves of Haiti's Tiburon Peninsula.

Subgenus *Syrrhophus* (41 spp.) inhabits Mexico, Guatemala, and Belize, with three species in Texas, including the Rio Grande Chirping Frog (*E. (Sy.) campi*), which has spread east into Louisiana, and two Cuban species. Of the 223 species in the subfamily, 68 are Critically Endangered and 67 Endangered.

DIET
Small invertebrates

IUCN STATUS
CR = 68, EN = 67, VU = 19, NT = 9; percentage species in trouble = 75%

BRIGHT-EYED MALAGASY TREEFROGS

The Mantellidae is the dominant Malagasy frog family, with three subfamilies. Boophinae comprises just the genus *Boophis*, and two subgenera, *Boophis* (70 spp.), which inhabit rainforests and breed in streams, and *Sahona* (10 spp.), from open habitats, breeding in small ponds. Numerous small eggs are attached to rocks or vegetation and tadpoles are free-swimming and exotrophic.

Small, nocturnal, and arboreal, bright-eyed treefrogs are distributed throughout Madagascar, in all habitats including plantations, but montane species may be terrestrial above the tree line.

A single species, the Mayotte Bright-eyed Treefrog (*B.* (*Sahona*) *nauticus*), is endemic to Mayotte in the Comoros Archipelago.

Boophis treefrogs have fingers and toes terminating in round discs, and little or no webbing. Their eyes have brightly colored irises and horizontal pupils. Most are cryptically patterned and several are polymorphic, with cryptic colors dominating. Some have translucent skin similar to the Neotropical glass frogs (Centrolenidae, see page 102).

Most are small but a number of species achieve over 2¾ in (70 mm), for example, the White-lipped Bright-eyed Treefrog (*B.* (*Boophis*) *albilabris*), Eastern Bright-eyed Treefrog (*B.* (*S.*) *opisthodon*), and Goudot's Bright-eyed Treefrog (*B.* (*B.*) *goudotii*).

Five species are listed by the IUCN as Critically Endangered, 18 Endangered, 12 Vulnerable, and 6 Near Threatened.

LEFT | The White-lipped Bright-eyed Treefrog (*Boophis albilabris*) is a large species distributed widely in the north and east of Madagascar.

OPPOSITE | The Malagasy Bullfrog (*Laliostoma labrosum*) is a large and widely distributed species from western Madagascar.

DISTRIBUTION
Madagascar, Mayotte

GENUS
Boophis

HABITATS
Primary and secondary rainforest streams, banana plantations, and occasionally montane forest

SIZE
¾ in (21 mm) ♂ Liam's Bright-eyed Malagasy Treefrog (*B.* (*Boophis*) *liami*) to 3½ in (87 mm) ♀ Goudot's Bright-eyed Malagasy Treefrog (*B.* (*B.*) *goudoti*)

ACTIVITY
Nocturnal; arboreal, semiterrestrial

REPRODUCTION
Axillary amplexus; 200–400 small eggs are laid in streams or stagnant ponds

DIET
Presumed forest floor invertebrates

IUCN STATUS
CR = 5, EN = 19, VU =12, NT = 6; percentage species in trouble = 51%

MALAGASY BULLFROG & MALAGASY JUMPING FROGS

Laliostominae is a small subfamily containing two genera, both endemic to Madagascar. The nominate genus *Laliostoma* is monotypic, containing the Malagasy Bullfrog (*L. labrosum*), a short-headed, stocky-bodied, open, arid habitat-dwelling frog. Variously patterned unicolor or blotched brown, it is terrestrial but adopts a semi-fossorial existence in dry weather, only to emerge during the first heavy rains to breed explosively in watercourses, including rice paddies. It is absent from the rainforests of eastern Madagascar.

Aglyptodactylus contains six species of Malagasy jumping frogs. Relatively stout, ranid-like with long, powerful legs, the jumping frogs are terrestrial inhabitants of leaf-litter in both dry deciduous western forests and wet eastern rainforests. They occur throughout Madagascar but favor coastal locations. The jumping frogs are also explosive breeders that appear in numbers during the wet season to breed in any temporary ponds. Eggs are laid in the water and the tadpoles are free-swimming and exotrophic.

Jumping frogs are of moderate size, the smallest being the male Western Malagasy Jumping Frog (*A. securifer*), which is larger than most *Boophis*. One species each are listed as Critically Endangered, Vulnerable, and Near Threatened.

DISTRIBUTION
Madagascar

GENERA
Aglyptodactylus, Laliostoma

HABITATS
Open habitats, rice paddies, arid deciduous forest, and rainforest

SIZE
1½ in (35 mm) ♂ Western Malagasy Jumping Frog (*A. securifer*) to 2½ in (64 mm) ♀ Malagasy Bullfrog (*L. labrosum*)

ACTIVITY
Diurnal; terrestrial, semi-fossorial

REPRODUCTION
Axillary amplexus; explosive breeders in temporal ponds after heavy rain

DIET
Presumed forest floor invertebrates

IUCN STATUS
EN = 1, VU =1, NT = 1; percentage species in trouble = 43%

MANTELLAS & MALAGASY LITTER, PANDANUS & TREEFROGS

BELOW | The Eastern Malagasy Frog (*Gephyromantis plicifer*) is a common species in the lowland wet forests.

Mantellinae is the largest mantellid subfamily comprising 9 genera and 151 species, all but one endemic to Madagascar and its satellite islands such as Nosy Be. The largest genera are *Gephyromantis* (51 spp.) and *Mantidactylus* (37 spp.), Malagasy litter frogs which are primarily, though not exclusively, terrestrial rainforests inhabitants. Some species lay their eggs close to water with the result that their tadpoles are free-swimming, but many other species are direct breeders, laying eggs in the leaf-litter, which hatch into froglets. *Blommersia* (12 spp.) contains terrestrial direct-breeding but semiarboreal species that lay their eggs in vegetation over water. One *Blommersia* is endemic to Mayotte (*Bl. transmarina*).

The most famous genus is *Mantella* (16 spp.), the Madagascan poison frogs or mantellas, which demonstrate convergence with the Neotropical dart-poison frogs (Dendrobatidae, see pages

LEFT | Variegated Mantellas (*Mantella baroni*) are black, red, and green above and black with light blue spots below.

OPPOSITE | A small species, the Tsarafidy Frog (*Guibemantis pulcher*) inhabits the leaf-axils of screw-pines (*Pandanus*).

DISTRIBUTION
Madagascar and Mayotte

GENERA
Blommersia, Boehmantis, Gephyromantis, Guibemantis, Mantella, Mantidactylus, Spinomantis, Tsingymantis, Wakea

HABITATS
Rainforest ponds, swamps, streams, cocoa plantations, and limestone karst (*Tsingymantis*)

SIZE
½ in (13 mm) ♂ Sambirano Dwarf Malagasy Frog (*W. madinika*) to 4¾ in (120 mm) Gray Madagascan Frog (*Mantidactylus guttulatus*)

ACTIVITY
Nocturnal and arboreal (*Blommersia, Guibemantis, Spinomantis*), or diurnal and terrestrial (*Mantella*)

REPRODUCTION
Eggs laid on leaves over water (*Blommersia, Guibemantis*); free-swimming tadpoles or direct breeders (*Gephyromantis*)

117–120), sequestering toxins from their invertebrate diet and advertising their toxicity through aposomatic patterning. The Golden Mantella (*Mantella aurantiaca*) is almost indistinguishable from the Golden Dart-poison Frog (*Phyllobates terribilis*), while other mantellas are also brightly colored, such as the red and black Cowen's Mantella (*Mantella cowanii*) or yellow and black Variegated Mantella (*M. baroni*), but some species are drab and cryptic.

DIET
Presumed forest floor
invertebrates

IUCN STATUS
CR = 7, EN = 25, VU =24,
NT = 8; percentage species
in trouble = 42%

Guibemantis and *Spinomantis* contain 18 and 14 treefrog species respectively, several of the former genus being associated with *Pandanus* screw-pine. *Spinomantis* is named for the distinctive toothlike fringes that adorn the legs of some species.

Mantellinae also contains three monotypic species. The Malagasy Limestone Frog (*Tsingymantis antitra*) is a large species from extreme northern Madagascar where it inhabits karst outcrops. The Sambirano Dwarf Frog (*Wakea madinika*) is the smallest mantellid, while Böhme's Malagasy Frog (*Boehmantis microtympanum*) is at the other end of the scale, one of the largest species.

Seven species are Critically Endangered, including the Black-eared Mantella (*Mantella milotympanum*), 25 are Endangered, including the Malagasy Limestone Frog, 24 are Vulnerable, and 8 are Near Threatened.

KAJIKA & EAST ASIAN STREAM FROGS

TOP | Found throughout the Japanese Ryukyu Islands, the Ryukyu Kajika Frog (*Buergeria japonica*) is a small but common species.

ABOVE | The Robust Stream Frog (*Buergeria robusta*) is a large species that is endemic to Taiwan.

The Rhacophoridae contains two subfamilies. Buergeriinae comprises a single genus, *Buergeria* (6 spp.) of smooth-skinned stream frogs. The northernmost species is the Japanese Kajika Frog (*Buergeria buergeri*), from Japan's main islands south of Hokkaidō. Kajika means "river deer" and is a reference to the call of males, which sounds like "ojika," which translates as "buck." Kajika is also a name for someone with a singing voice, and kajika frogs are kept in captivity for their calling.

The Ryukyu Kajika Frog (*B. japonica*) and the Yaeyama Kajika Frog (*B. choui*) inhabit the Ryukyus and Taiwan. Taiwan is also home to the Robust Stream Frog (*B. robusta*) and Ota's Stream Frog (*B. otai*). The southernmost species, the Hainan Stream Frog (*B. oxycephala*), occurs between 260–2,620 ft (80–800 m) elevation, and is listed as Vulnerable, being threatened by habitat loss even though it often survives in severely degraded forests.

These frogs have fully webbed toes and half-webbed fingers. They live along forest streams from coastal to montane localities, where males call from rocks, often using the same rock for months. Eggs are laid individually or in small clumps, the developing tadpoles being strong swimmers.

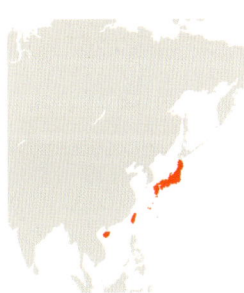

DISTRIBUTION
Japan, Ryukyu Islands, Taiwan, and Hainan Island

GENUS
Buergeria

HABITATS
Forest streams, coastal lowland, highland, and degraded forest

SIZE
1¼ in (29 mm) Yaeyama Kajika Frog (*B. choui*) to 3¼ in (85 mm) Japanese Kajika Frog (*B. buergeri*)

ACTIVITY
Nocturnal; terrestrial, arboreal, aquatic

REPRODUCTION
Axillary amplexus; the eggs are laid in the water and pass

through a tadpole stage in two months

DIET
Insects and spiders

IUCN STATUS
VU = 1; percentage species in trouble = 16%

AFRICAN FOAM-NEST TREEFROGS

Rhacophorinae is a large subfamily containing 22 genera and 440 species. It is primarily Asian, but one genus, *Chiromantis* (4 spp.), contains the African foam-nest treefrogs. The Southern Foam-nest Frog (*C. xerampelina*) inhabits East Africa south to Eswatini and Zululand. In Kenya and Tanzania it overlaps with Peters' Foam-nest Frog (*C. petersii*), and in northern Kenya and Somalia with Keller's Foam-nest Frog (*C. kelleri*). The Western Foam-nest Frog (*C. rufescens*), occurs from Uganda to West Africa, including Bioko Island, in the Gulf of Guinea. Habitat preferences include arid savanna (*C. kelleri, C. petersi*), savanna-woodland (*C. xerampelina*), and rainforest (*C. rufescens*).

These frogs have large eyes, horizontal pupils, webbed fingers and toes with large round discs, and they are excellent climbers. Their rugose skin is colored gray, brown, and green like tree bark. They sleep with their legs tucked alongside the body to conserve water, and break up their outline. A pair produce a foam-nest by using their legs to beat the fluid produced with the eggs. Nests are produced in vegetation overhanging water where the eggs are protected once the foam hardens, but they are still vulnerable to monkeys, which eat the foam, and Fornasini's Spiny Reedfrog (*Afrixalus fornasini*), which burrows into the nest to eat the eggs.

BELOW | The Southern Foam-nest Treefrog (*Chiromantis xerampelina*) is a highly arboreal inhabitant of East and Southern Africa.

DISTRIBUTION South and Southeast Asia, also sub-Saharan Africa **GENERA** *Beddomixalus, Chirixalus, Chiromantis, Feihyla, Ghatixalus, Gracixalus, Kurixalus, Leptomantis, Mercurana, Nasutixalus, Nyctixalus, Philautus, Polypedates, Pseudophilautus,*	*Raorchestes, Rhacophorus, Rohanixalus, Romerus, Taruga, Theloderma, Vampyrius, Zhangixalus* **HABITATS** Forest, woodland, savanna, and plantations **SIZE** ¾ in (18 mm) Lanjak Bushfrog (*Ph. rufugii*) to 4 in (102 mm) Denny's Treefrog (*Z. dennysi*)	**ACTIVITY** Nocturnal; arboreal, terrestrial **REPRODUCTION** Eggs in foam-nests or tree holes hatch into tadpoles or froglets **DIET** Small invertebrates **IUCN STATUS** EX = 17, CR = 31, EN = 62, VU = 39, NT = 18; percentage species in trouble = 38%

ASIAN FOAM-NEST TREEFROGS

The Asian rhacophorine frogs occur throughout Asia in most habitats, with some of the highest diversity in Sri Lanka and southern India. The Common Indian Whipping Treefrog (*Polypedates maculatus*) occurs to 9,840 ft (3,000 m) elevation in the Himalayas, while Hu's Treefrog (*Zhangixalus hui*) is found at 10,330 ft (3,150 m) in Sichuan, China.

The genera *Chirixalus*, *Feihyla*, *Ghatixalus*, *Polypedates*, *Rhacophorus*, and *Rohanixalus* are variously referred to as foam-nest frogs, jelly-nest frogs, bubble-nest frogs, or whipping frogs, because of their reproductive strategy of whipping up a foam-nest, like *Chiromantis* in Africa. When the eggs hatch the tadpoles drop into the water where they will feed and develop. Not all species build nests; the ridge-nosed treefrogs (*Nyctixalus*) and warty or bug-eyed treefrogs (*Theloderma*) lay their eggs in tree holes where they hatch into nonfeeding tadpoles that develop rapidly into froglets, while Oriental shrub frogs (*Philautus*) and Sri Lankan shrub frogs (*Pseudophilautus*) use direct development, laying eggs on the ground that hatch into small froglets, or attaching their eggs to leaves overhead. The female Taiwan Frilled Treefrog (*Kurixalus eiffingeri*) lays her eggs in water-filled bamboo stalks and periodically lays infertile eggs to feed her oophagous tadpoles.

BELOW | Wallace's Flying Frog (*Rhacophorus nigropalmatus*) glides down from the canopy by using the webbing on its feet like a parachute.

Members in the nominate genus *Rhacophorus*, and a few species in other genera, are referred to as flying frogs because they escape predators by gliding or parachuting to the ground by spreading the large webs on their feet and flattening their bodies. The most famous is Wallace's Flying Frog (*R. nigropalmatus*), first reported by the nineteenth-century British naturalist Alfred Russel Wallace. It has huge outsized webbed hands and feet and is one of the largest rhacophorids.

Many Asian rhacophorine frogs are threatened, for example, 31 are Critically Endangered, including the Resplendent Bushfrog (*Raorchestes resplendens*) of southern India, 62 are Endangered, 39 Vulnerable, and 18 Near Threatened. The most endangered genus is *Pseudophilautus* from Sri Lanka and southern India; only 6 of its 80 species are not considered threatened and 17 are already believed Extinct.

TOP | From western Java, the Javan Jelly-nest Treefrog (*Feihyla vittiger*) is a small species that also inhabits tea plantations.

INSET | The Western Ghats Foam-nest Frog (*Ghatixalus asterops*) inhabits one of the richest biodiversity hotspots in the world.

AFRICAN RIVER, DAINTY & SAND FROGS

BELOW LEFT | Mountain Dainty Frogs (*Caocosternum parvum*) inhabit high elevation grasslands in eastern South Africa.

BELOW | The Cryptic Sand Frog (*Tomopterna cryptotis*) is one of 15 rotund and arid-adapted frogs found across East and Southern Africa.

OPPOSITE | Delaland's River Frog (*Amietia delalandii*) is a large, powerful frog associated with water courses from Zambia to the Cape.

Cacosterninae contains ten genera. *Amietia* (16 spp.) are river frogs distributed from Ethiopia to South Africa, in many aquatic habitats. They have powerful hindlimbs, pointed snouts, and large eyes. Although generally nocturnal, De Witte's River Frog (*Am. wittei*) lives at high elevations (6,560–9,840 ft [2,000–3,000 m]) on Mount Kilimanjaro and basks during the day. The high-elevation Maluti River Frog (*Am. vertebralis*) overwinters under ice and eats crabs, while the Angolan River Frog (*Am. angolensis*) can leap to catch flying insects.

Strongylopus (10 spp.) are stream frogs found from Tanzania to South Africa. They are smaller and more gracile that river frogs, and often striped. They lay eggs in stream-side vegetation. The dainty frogs, *Cacosternum* (16 spp.) from Ethiopia to South Africa, are also small and occur from high-elevation grassland to coastal fynbos, breeding in temporary ponds. Most are smooth-skinned but the Cape Dainty Frog (*C. capense*) has wartlike tubercles on its dorsum.

DISTRIBUTION
Sub-Saharan Africa

GENERA
Amietia, Anhydrophryne, Arthroleptella, Cacosternum, Microbatrachella, Natalobatrachus, Nothophryne, Poyntonia, Strongylopus, Tomopterna

HABITATS
Forest, savanna woodland, montane forest, grassland, sandy areas, rocky slopes, swamps, ponds, rivers, streams, fynbos, and agricultural land to 10,825 ft (3,300 m)

SIZE
½ in (12 mm) ♂ Smooth Dainty Frog (*C. platys*) or ♂ Striped Dainty Frog (*C. striatum*) to 6 in (150 mm) ♀ Maluti River Frog (*Am. vertebralis*)

ACTIVITY
Diurnal, nocturnal; terrestrial, aquatic, semi-fossorial, semiarboreal

The semi-fossorial sand frogs, *Tomopterna* (16 spp.), which inhabit dry habitats from Mauritania to Somalia and south to the Cape, are large, short-legged frogs. The small, terrestrial, mongrel frogs, *Nothophryne* (5 spp.), are endemic to the granite inselbergs of Malawi and Mozambique. Quirimbas Mongrel Frog (*No. unilurio*) is Critically Endangered, the other four are Endangered.

Several endemic cacosternine genera inhabit South Africa, for example, the tiny direct-developing moss frogs, *Arthroleptella* (10 spp.) of Western Cape, with the Rough Moss Frog (*Ar. rugosa*) and Quiet Moss Frog (*Ar. subvoce*) Critically Endangered. The same status is given to the monotypic Cape Micro Frog (*Microbatrachella capensis*), from the fynbos and marshes of Cape Town. The Cape Marsh Frog (*Poyntonia paludicola*) also occurs here and is listed as Near Threatened.

Another area of endemicity is the Eastern Cape–KwaZulu-Natal region. *Anhydrophryne* (3 spp.) are small, direct-developing, chirping frogs that inhabit streams. The Ngongoni Chirping Frog (*An. ngongoniensis*) is Endangered while the semiarboreal Boneberg's Kloof Frog (*Natalobatrachus bonebergi*) is Critically Endangered due to habitat loss to sugarcane production.

REPRODUCTION
Explosive breeders in wet season, eggs laid in water hatching into tadpoles, or nonfeeding tadpoles that rapidly become froglets, or eggs on land with direct development to froglets

DIET
Invertebrates, including flying insects and crabs

IUCN STATUS
CR = 4, EN = 9, VU = 2, NT = 7; percentage species in trouble = 23%

AFRICAN BULLFROGS & FISHING FROGS

BELOW | The Giant Bullfrog (*Pyxicephalus adspersus*) is in Africa what *Ceratophrys* are in South America, a voracious predator of smaller vertebrates.

OPPOSITE | Said to feed on small fish as they leap from the water, Brown Fishing Frogs (*Aubria subsigillata*) of West Africa are also hunted by villagers.

The Pyxicephalinae contains two genera, the most familiar being *Pyxicephalus* (5 spp.), the African bullfrogs, epitomized by the Giant Bullfrog (*P. adspersus*), a savanna-woodland inhabitant found from Kenya to South Africa. It is commonly kept in captivity where it is called the "pixie frog," though it is far from fairylike.

Females are large (up to 7 in [180 mm] SVL), but males are immense, achieving 9–9½ in (230–245 mm), a rare occasion where male anurans are larger than females. The Giant Bullfrog is olive green above, contrasting the white of its irises, and lemon yellow below, and its dorsum is covered in large wartlike tubercles and ridges.

DISTRIBUTION
Sub-Saharan Africa

GENERA
Aubria, Pyxicephalus

HABITATS
Savanna, savanna woodland, lowland rainforest, and swamps

SIZE
3¼ in (81 mm) ♂ Masako Fishing Frog (*A. masako*) to 9¾ in (245 mm) ♀ Giant Bullfrog (*P. adspersus*)

ACTIVITY
Nocturnal; terrestrial, semi-fossorial

REPRODUCTION
Explosive breeders in the wet season, eggs laid in water, tadpole stage

DIET
Fish (*Aubria*), invertebrates, frogs, reptiles, and mice (*Pyxicephalus*)

IUCN STATUS
Low Concern, no species are listed as under threat

The four smaller species inhabit arid habitats that experience seasonal heavy rain, the Edible Bullfrog (*P. edulis*) on the East African coast from Somalia to Mozambique and inland to Zambia, the Biera Bullfrog (*P. angusticeps*) from Tanzania to Mozambique, the Obbia Bullfrog (*P. obbianus*) in northern coastal Somalia, and an unnamed species in the arid Sahel from Senegal to Sudan.

Genus *Aubria* (2 spp.) contains the fishing frogs: the Brown Fishing Frog (*A. subsigillata*), from Liberia to Gabon, and the Masako Fishing Frog (*A. masako*) from Cameroon to the Democratic Republic of the Congo. Dorsally these frogs are olive green or reddish-brown, but their venters are distinctive, black with large white or yellow spots.

Bullfrogs are explosive breeders at the start of the rainy season when they congregate in large numbers in any available water. Females lay 3,000–4,000 eggs in temporary ponds where they are guarded by the attentive males that then construct channels to direct the tadpoles to more permanent water. Fishing frogs also lay numerous eggs and their tadpoles shoal together densely upon hatching.

African bullfrogs are voracious; they will swallow anything they can fit into their mouths, which are armed with a pair of mandibular tusks, including frogs, small snakes, and mice. The fishing frogs are believed to feed on small fish.

GIANT FROGS & SLIPPERY FROGS

Family Conrauidae contains a single genus, *Conraua* (8 spp.), which includes the world's largest anuran, the Goliath Frog (*C. goliath*). Females achieve up to 8¾ in (220 mm) SVL while males may grow to 13½ in (340 mm). The other species are much smaller and may be called "slippery frogs."

The Goliath Frog occurs from southern Cameroon to Equatorial Guinea up to elevations of 3,280 ft (1,000 m) in fast-flowing rainforest streams and rivers. Highly aquatic, it is stout-bodied with long, powerful hindlimbs and fully webbed toes. Males have a strange call that starts with a whistle and ends with a roar. They build large nests out of rocks and when a female arrives they engage in amplexus. She then lays 150–2,800 eggs in the nest, which she guards. Goliath Frogs rest on riverine rocks, escaping to deep water with one large bound if danger threatens. Unsurprisingly, such a large frog, which may weigh 2.2–8.8 lb (3–4 kg), is much sought after for the bushmeat trade, and this is pushing it to the brink of extinction and it is listed as Endangered. The Goliath Frog is large enough to include crabs, snails, fish, other frogs, young turtles, small snakes, and small mammals in its diet.

DISTRIBUTION
West, Central, and East Africa

GENUS
Conraua

HABITATS
Lowland and montane rainforest streams and rivers

SIZE
2½ in (61 mm) ♀ Allen's Giant Frog (*C. alleni*) to 13½ in (340 mm) ♂ Goliath Frog (*C. goliath*)

ACTIVITY
Usually nocturnal; aquatic, semiterrestrial

REPRODUCTION
Males build a nest, females lay up to 2,800 eggs that they then guard; tadpoles are herbivorous

DIET
Invertebrates, amphibians, fish, small snakes, and mammals

IUCN STATUS
CR = 2, EN = 1, VU = 2; percentage species in trouble = 63%

The Thick-legged Slippery Frog (C. *crassipes*) occurs from Nigeria to the Republic of the Congo, the Angolan exclave of Cabinda, and Bioko Island, Gulf of Guinea. It is not considered endangered, but the Robust Giant Frog (C. *robusta*), which has a smaller range than the Goliath Frog on the Cameron–Nigeria border, is considered Vulnerable.

Four species of *Conraua* inhabit West Africa from Guinea to Togo and two of these, De Roo's Slippery Frog (C. *derooi*) in Ghana and Togo, and the recently described Atewa Slippery Frog (C. *sagyimase*) from Ghana, are Critically Endangered. Beccari's Giant Frog (C. *beccari*), inhabits the highlands of Ethiopia and Eritrea at 2,625–8,200 ft (800–2,500 m) elevation.

AFRICAN TORRENT FROGS & BALE MOUNTAIN FROG

Authors differ over the composition of the Petropedetidae. As recognized it contains three genera: the Central and East African torrent frogs, *Petropedetes* (9 spp.) and *Arthroleptides* (3 spp.), and the monotypic Bale Mountain Frog, *Ericabatrachus baleensis*.

The torrent frogs occur in similar habitats, splash zones of rocky rainforest streams, but on opposite sides of the continent. All nine Central African species occur in southern Cameroon, with several extending west into Nigeria, south to Gabon, or onto Bioko Island (3 spp.). The Cameroon Torrent Frog (*P. cameronensis*) occurs up to 4,590 ft (1,400 m) elevation. These are slender to stout frogs with rugose skin, long legs, and digits terminating in bilobed discs, webbing varying from minimal to complete, and sexually dimorphic tympana— smaller than the eyes in females, larger in males. Eggs are laid on wet rocks, and are guarded by the

LEFT | Parker's Torrent Frog (*Petropedetes parkeri*) inhabits rocky areas alongside fast-flowing streams, up to 3,200 ft (1,000 m) elevation.

OPPOSITE | The Endangered Southern Torrent Frog (*Arthroleptides yakusini*) exhibits a scattered distribution across Tanzania.

DISTRIBUTION
Central and East Africa

GENERA
Arthroleptidae, Ericabatrachus, Petropedetes

HABITATS
Lowland and montane rainforest streams, also high-elevation heath woodland

SIZE
¾ in (22 mm) ♂ Bale Mountain Frog (*E. baleensis*) to 2¾ in (73 mm) ♂ Southern Torrent Frog (*A. yakusini*)

ACTIVITY
Nocturnal; terrestrial, saxicolous

REPRODUCTION
Eggs laid on wet rocks or in leaf-litter; some males guard eggs

DIET
Presumed small invertebrates

IUCN STATUS
CR = 9, EN = 5, VU = 3, NT = 5; percentage species in trouble = 23%

males, while tadpoles live in the splash zone. Parker's Torrent Frog (*P. parkeri*) and Johnston's Torrent Frog (*P. johnstoni*) lay eggs in leaf-litter, often far from water. Perret's Torrent Frog (*P. perreti*) and the White-spotted Torrent Frog (*P. palmipes*) are Endangered.

The East African torrent frogs are Martienssen's Torrent Frog (*A. martiensseni*) and the Southern Torrent Frog (*A. yakusini*) from Tanzania, and the Mount Elgon Torrent Frog (*A. dutoiti*), which is found to 7,220 ft (2,200 m) elevation in Kenya. Morphologically similar to *Petropedetes*, their tympana are much smaller than their eyes. The Tanzanian species are Endangered, while the Mount Elgon Torrent Frog is Critically Endangered, and having not been seen for many years, is feared extinct.

The Critically Endangered Bale Mountain Frog (*E. baleensis*) inhabits the Bale Mountains of Ethiopia, an area of high endemicity, from where the large Bale Mountain Adder (*Bitis harenna*) was described in 2016. The smallest petropedetid, it is a small green or brown frog with small dorsal warts, long unwebbed fingers and toes with bilobed discs. It is terrestrial in highland heath and forest, where females lay their eggs in the leaf-litter.

AFRICAN PUDDLE FROGS

Male

Female

ABOVE | The Bibita Mountain Dwarf Puddle Frog (*Phrynobatrachus bibita*), from Ethiopia, was only described in 2019.

The Phrynobatrachidae contains the single genus, *Phrynobatrachus* (95 spp.), the African puddle frogs. They lack webbing from their fingers, and most are small, but a few species are larger, for example, the Rough Puddle Frog (*P. asper*). *Phrynobatrachus* occurs widely throughout sub-Saharan Africa, excluding the Horn of Africa and the deserts of Namibia and Namaqualand, and also occur on the islands in the Gulf of Guinea, and Unguja (Zanzibar) and Pemba, Tanzania. They occur in habitats from dry savanna to humid montane evergreen forest up to 11,480 ft (3,500 m) elevation.

The common name "puddle frogs" comes from their ability to utilize almost any small, temporary water source for breeding, from rainforest rain puddles to cattle troughs and stagnant agricultural puddles. Some species are alternatively known as river frogs. Many species lay their eggs on the surface of the

DISTRIBUTION
Sub-Saharan Africa and Unguja Island

GENUS
Phrynobatrachus

HABITATS
Most habitats from dry savanna to evergreen monsoon forest, lowland to montane

SIZE
½ in (13 mm) ♂ Unguja Puddle Frog (*P. ungujae*) to 2¼ in (55 mm) Rough Puddle Frog (*P. asper*)

ACTIVITY
Diurnal, nocturnal; terrestrial

REPRODUCTION
Laying eggs on the water surface, or in tree holes

(*P. guineensis*) or leaf-litter (*P. phyllophilus, P. tokba*)

DIET
Insects, myriapods, and vegetation

IUCN STATUS
CR = 9, EN = 5, VU = 3, NT = 5; percentage species in trouble = 23%

water but the Guinea Puddle Frog (*P. guineensis*) uses tree holes, coconut husks, and even snail shells, while the Leaf-loving Puddle Frog (*P. phyllophilus*) and Tokba Puddle Frog (*P. tokba*) lay their eggs in humid leaf-litter.

As a genus, the member species demonstrate a wide variety of patterns, from cryptic browns to vivid greens, and from smooth to rugose skin, with the diurnal species exhibiting a degree of polymorphism that complicates species identification in an area with several species present.

Although puddle frogs are frequently encountered in sub-Saharan Africa, and the Natal Puddle Frog (*P. natalensis*) almost mirrors the generic range, it would be a mistake to think they are all common species. Most species have small ranges, and nine species are Critically Endangered, including the Lake Oku Puddle Frog (*P. njiomock*) and Spiny Puddle Frog (*P. chukuchuku*) from Mount Oku, Cameroon, and the Intermediate Puddle Frog (*P. intermedius*) of coastal Ghana; five are Endangered, including the smallest species, the Unguja Puddle Frog (*P. ungujae*) on Unguja Island (formerly Zanzibar), Tanzania, three are Vulnerable, and five Near Threatened.

SHARP-NOSED GRASS FROGS

Three genera comprise the Ptychadenidae, the largest being *Ptychadena* (64 spp.). These are the sharp-nosed grass frogs, moderately large frogs with pointed snouts, large eyes, and streamlined bodies with distinct longitudinal ridges, and powerful, webbed hindlimbs for leaping. There are exceptions—not all species have especially long snouts or distinctive ridges. While some species are uniform brown or blotched dorsally, most grass frog species exhibit a broad brown or green vertebral stripe, with finer pale dorsolateral stripes on the ridges to either side.

African grass frogs are terrestrial and inhabit a variety of habitats, from humid lowland rainforest to arid savanna woodland, and from high-elevation grasslands to farmed floodplains. They occur across sub-Saharan Africa, excluding the southwest and the extreme northeast. In the Gulf of Guinea, Bioko Island and São Tomé and Príncipe are inhabited by the Limbé Grass Frog (*P. aequiplicata*) and Newton's Grass Frog (*P. newtoni*) respectively. The Mascarene Grass Frog (*P. mascariensis*) inhabits

LEFT | The widely distributed Mascarene Grass Frog (*Ptychadena mascarensis*) is believed to be a species complex containing multiple species.

DISTRIBUTION
Sub-Saharan Africa and Nilotic Egypt

GENERA
Hildebrandtia, Lanzarana, Ptychadena

HABITATS
Humid and dry savanna, savanna woodland, marshes, lowland rainforest, high-elevation grassland, and rice paddy and agricultural habitats to 12,470 ft (3,800 m)

SIZE
1 in (27 mm) ♂ Smallest Grass Frog (*P. nana*) to 3¼ in (86 mm) Newton's Grass Frog (*P. newtoni*)

ACTIVITY
Nocturnal, diurnal; terrestrial, fossorial

REPRODUCTION
Axillary amplexus; breed in temporary pools, tadpoles often developing rapidly

DIET
Large invertebrates, e.g., beetles and crickets

IUCN STATUS
EN = 2, NT = 3; percentage species in trouble = 7%

Madagascar, the Seychelles, and Mascarene Islands (Mauritius), while the Nile Grass Frog (*P. nilotica*) is found along the length of the River Nile in Egypt, and qualifies as Mediterranean.

Most species lay their eggs in water, where development to larval and then froglet stages is fairly rapid, but Broadley's Grass Frog (*P. broadleyi*) lays eggs on wet rocks, and when the tadpoles hatch they remain in the wet film on the rocks.

The genus *Hildebrandtia* (3 spp.) contains the ornate frogs, stocky nocturnal species that inhabit arid savannas and spend much of the year underground, emerging to feed and breed in the wet season. Most widely distributed is the Common Ornate Frog (*H. ornata*), from the West African Sahel to East Africa and south to Angola and Botswana. The other two species are endemic to Somalia and Angola. The monotypic Lanza's Frog (*Lanzarana largeni*) is also an arid savanna dweller from Somalia.

Newton's Grass Frog and the Smallest Grass Frog (*P. nana*) are Endangered, while three species are Near Threatened.

TOP | A pair of Anchieta's Grass Frogs (*Ptychadena anchietae*) in amplexus.

INSET | The Common Ornate Frog (*Hildebrandtia ornata*) inhabits arid habitats and spends much of its life underground.

WESTERN HEMISPHERE LEOPARD, WATER & BROWN FROGS

RIGHT | The voracious American Bullfrog (*Aquarana catesbeiana*) occurs from Canada to Mexico but is invasive in many other parts of the world.

The Ranidae are the "true frogs," a large global family with 29 genera and over 430 species. The Americas contain 4 genera and over 60 species, the largest genus being *Lithobates* (57 spp.), which used to include species now placed in *Aquarana*. The largest American ranid is the American Bullfrog (*Aquarana catesbeiana*), which occurs from Canada to southern Mexico. Introduced widely, to Hawaii, Latin America, the Caribbean, Europe, and Asia, it is a threat to native vertebrates because of its size and voracious appetite. Other notable species include the Crayfish Frog (*L. areolatus*), which inhabits abandoned crayfish burrows, and the Pig Frog (*Aq. grylio*) of Florida, which has a grunting call. The numerous leopard frogs, which have long jumping legs, strongly webbed toes, and bold spotted patterns, include the Northern and the Southern Leopard Frog (*L. pipiens* and *L. sphenocephalus*).

The Wood Frog (*Boreorana sylvatica*) inhabits woodlands and bogs across northern Canada and Alaska. It is the most cold-adapted amphibian in the world and has been allocated its own genus. During

DISTRIBUTION
Asia, Europe, Africa, Americas, and Australasia

GENERA
Abavorana, Amerana, Amnirana, Amolops, Aquarana, Babina, Boreorana, Chalcorana, Clinotarsus, Glandirana, Huia, Humerana, Hydrophylax, Hylarana, Indosylvirana, Lithobates, Meristogenys, Nidirana, Odorrana, Papurana, Pelophylax,
Pseudorana, Pterorana, Pulchrana, Rana, Sanguirana, Staurois, Sumaterana, Sylvirana, Wijayarana

HABITATS
Rainforest, woodland, grassland, mountains, wetlands, rivers, streams, lakes, ponds, and caves

SIZE
1 in (25 mm) ♂ Lesser Splash Frog (*St. parvus*) to 8¾ in (220 mm) American Bullfrog (*Aq. catesbeiana*)

hibernation its body temperature will drop below the freezing point of water, and it will freeze, but glucose that floods its circulatory system prevents any frost damage, so it thaws again in the spring.

From Mexico, *Lithobates* ranges south in northern South America but with far fewer species, only three reaching Ecuador, and just one, the widespread Amazon Water Frog (*L. palmipes*) in Amazonia.

Five *Lithobates* are Critically Endangered, for example, the Nicaraguan Little Corn Island Frog (*L. miadis*), the high-elevation (7,624 ft [2,324 m])

Lago de las Minas Frog (*L. chichicuahutla*), Puebla Frog (*L. pueblae*), Tlaloc's Leopard Frog (*L. tlaloci*), from Mexico, and the Dusky Gopher Frog (*L. sevosus*) from Louisiana.

The newly erected genus *Amerana* inhabits North America, from British Columbia to Baja California, where *Lithobates* is absent. Most of the eight species are referred to as red-legged or yellow-legged frogs, for example, the Northern Red-legged Frog (*A. aurora*). The northernmost species, the Columbia Spotted Frog (*A. luteiventris*) reaches southeastern Alaska.

ACTIVITY
Diurnal, nocturnal; terrestrial, aquatic, arboreal

REPRODUCTION
From still water ponds to fast-flowing streams, laying egg clumps; some tadpoles have belly suckers, most metamorphose in months, though *Aq. catesbeiana* may take 1–2 years to become a froglet

DIET
Mostly invertebrates, but some species, e.g., *Aq. catesbeiana*, are large enough to prey on other frogs and small vertebrates, even snakes

IUCN STATUS
CR = 9, EN = 33, VU = 43, NT = 24; percentage species in trouble = 25%

ABOVE | After breeding, Columbia Spotted Frogs (*Amerana luteiventris*) move away from their breeding sites to avoid predation by gartersnakes (*Thamnophis* spp.) that arrive to feed on the tadpoles.

EASTERN HEMISPHERE BROWN, WATER & CASCADE FROGS

The Ranidae is a dominant family in Eurasia and Tropical Asia, with 25 genera and 360 species. The brown frog genus *Rana*, encountered in western North America, is one of the large Palearctic genera with 47 Eurasian species. The Common Frog (*R. temporaria*), is found from Ireland to Kazakhstan. Many of the Palearctic *Rana* are confusingly similar, being brown, with powerful hind limbs, strongly webbed toes, large eyes, a dark stripe from the snout through the eye and tympanum to the shoulder, and a raised dorsolateral fold along each side of the body.

The water frogs, *Pelophylax* (22 spp.), are usually green and while brown *Rana* prefer cool shady watercourses, they are sun-lovers. Brown frog species may be distinguished apart by hindleg length, but some water frogs require DNA analysis to separate them, especially as there are hybrid species created by mating between different species, for example, the Marsh Frog (*Pe. ridibundus*) and the Iberian Water Frog (*Pe. perezi*) in southern France and northern Spain, which resulted in Graf's Hybrid Frog (*Pe. kl. graft*) and the Marsh Frog and Pool Frog (*Pe. lessonae*), which resulted in the European Edible Frog (*Pe. kl. esculentus*). The "kl." means this is a "klepton species," a species produced by hybridogenesis.

The largest ranid genus is *Amolops* (73 spp.), the torrent or cascade frogs found from Nepal and India to China and Malaysia, though torrent- or cascade-dwelling frogs are also present in other genera, for example, *Sumaterana* (3 spp.) in Sumatra,

OPPOSITE | The Common Frog (*Rana temporaria*) is a widespread species through western and northern Europe.

ABOVE | Known from only two locations in the United Kingdom, the European Pool Frog (*Pelophylax lessonae*) is widely distributed in Europe.

RIGHT | Widely distributed in Borneo, the Black-spotted Splash Frog (*Staurois guttatus*) is semiarboreal along rocky streams.

Wijayarana (5 spp.) in Sumatra, Java, and Thailand, and the
splash frogs (*Staurois*, 6 spp.) in Borneo and the Philippines.
All these frogs are associated with fast-running, often rocky
rainforest or montane streams.

Amolops have flattened bodies to allow water to pass over
them, and finger and toe pads to grip the rocks, and their
tadpoles have large suckers on their bellies for the same
purpose. The Rufous-spotted Torrent Frog (*Amolops loloensis*)
occurs up to 12,140 ft (3,700 m) in Sichuan and Yunnan in
China. The Bornean and Philippine *Staurois*, such as the
attractive Black-spotted Splash Frog (*St. guttatus*), are also known
as "foot-flagging frogs" because the males perform elaborate
displays of hindfoot waving to deter other males and attract
females. The sound of the water would drown out any calls.

Odorrana (62 spp.) contains aquatic cascade-dwelling frogs
found from India to China, Borneo, and Japan. Hose's Rock
Frog (*O. hosii*) inhabits forest streams from Myanmar to

Borneo. Its skin contains foul-smelling toxins that can kill small vertebrates. A Chinese *Odorrana* has been reported to inhabit caves. The Chinese Concave-eared Frog (*O. tormota*) has an auditory canal like a mammal, with the tympanum deep inside, while most anurans have external tympana. The same strange characteristic is known for *Huia cavitympanum*, the Borneo Hole-in-the-head Frog (see page 23), a large frog found along fast-flowing streams in hill forests. These two species call using ultrasound, and it is believed the recessed tympanum improves their hearing.

The Australasian-Melanesian genus *Papurana* (19 spp.) includes the large Arfak River Frog (*Pa. arfaki*) from New Guinea and while other species occur in the Lesser Sunda Islands, the Moluccas and mainland Southeast Asia, the genus is absent from the Greater Sundas. A single species occurs in the Solomon Islands, the San Cristoval Frog (*Pa. kreffti*), and another species is found in northern Australia, the Australian Wood Frog (*Pa. daemeli*), which is misnamed, because the greater part of its range lies in New Guinea.

Pelophylax is also found in North Africa with the Saharan Green Frog (*Pe. saharicus*) inhabiting semidesert and mountains between the Mediterranean and the Sahara. The Levantine Green Frog (*Pe. bedriagae*) inhabits Turkey and the Levant. *Amnirana* (13 spp.) is the only sub-Saharan ranid genus, containing the white-lipped frogs of West, Central, and East Africa, such as the Long-legged White-lipped Frog (*Amnirana longipes*).

Nine ranid species are Critically Endangered (5 in *Lithobates*), 33 Endangered, 43 Vulnerable, and 24 Near Threatened.

ALCALA'S MOUNTAIN FROGS & PAPILLA-TONGUED FROGS

LEFT | The Borneo Dwarf Mountain Frog (*Alcalus baluensis*) has a scattered distribution throughout Borneo, but it is poorly known in nature.

OPPOSITE | The Himalayan Papilla-tongued Frog (*L. himalayana*) is only known from the high-elevation Tale Valley Wildlife Sanctuary in Arunachal Pradesh, India. It was only described in 2019.

Ceratobatrachidae is a diverse family containing three subfamilies, two of which comprise single genera, while the other contains two genera. One characteristic that all ceratobatrachids (for which the reproductive strategy is known) have in common is direct breeding—they lay a small number of large eggs in the leaf-litter that hatch into froglets, there is no tadpole stage. Otherwise they exhibit a wide variety of body forms and ecological preferences.

The subfamily Alcalinae contains five species of mountain frogs in the genus *Alcalus*, named for the Filipino biologist Angel Alcala. Three species inhabit Borneo, one occurs on the Philippine island of Palawan, and the fifth species occurs on the Southeast Asian mainland, in southern Myanmar

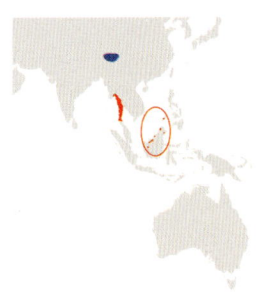

DISTRIBUTION
(1) Myanmar, Thailand, Malaysia, Borneo, and the Philippines (Palawan), (2) China (Xizang/Tibet) and northeast India

GENERA
(1) *Alcalus*; (2) *Liurana*

HABITATS
Montane rainforest streams, and mixed montane evergreen or temperate forest

SIZE
1 in (25 mm) ♂ Dwarf Mountain Frog (*A. baluensis*) to 1¾ in (43 mm) ♀ San Cascade Frog (*A. tasanae*); ½ in (14.2 mm) ♂ Minute Papilla-tongued Frog (*L. minuta*)

to 1¼ in (30.5 mm) ♀ Xizang Papilla-tongued Frog (*L. xizangensis*)

ACTIVITY
Nocturnal; terrestrial

REPRODUCTION
(1) Little-known, probably direct breeding in leaf-litter; (2) direct development, eggs laid in damp leaf-litter, hatching into froglets

Key: (1) Alcalinae (red), (2) Liuraninae (blue)

and peninsular Thailand. All are relatively small but stockily-built, terrestrial, montane rainforest frogs with extremely localized distributions. They have rough dorsal skin and are cryptically patterned. The Palawan Mountain Frog (*A. mariae*) and Saibau Mountain Frog (*A. sariba*) are Endangered.

The second subfamily, Liuraninae, is named for the Chinese herpetologist Ch'eng-chao Liu, and it occurs a considerable distance to the north of the other two subfamilies, in the Himalayas. *Liurana* comprises seven species, which have been referred to as papilla-tongued frogs because of the presence of lingual papillae. Small and cryptically patterned, the distribution of these montane evergreen and temperate forest leaf-litter dwelling frogs is centered in the Medog region of China, in Xizang (Tibet), and Arunachal Pradesh, northeastern India. They have been recorded at 1,800–10,500 ft (550–3,200 m) elevation. The Alpine Papilla-tongued Frog (*L. alpina*) and the Xizang Papilla-tongued Frog (*L. xizangensis*) are Vulnerable.

DIET
Presumed small forest floor invertebrates

IUCN STATUS
EN = 2, VU = 2; NT = 1; percentage species in trouble = 42%

MELANESIAN FROGS

LEFT | Günther's Triangle Frog (*Cornufer guentheri*), from the Solomon Islands, is an extremely variable species ranging from gray to bright orange.

The subfamily Ceratobatrachinae has undergone considerable taxonomic change in recent years but it is now considered to comprise just two genera, *Cornufer* (58 spp.) and *Platymantis* (32 spp.). *Platymantis* is endemic to the Philippines, excluding Palawan, while *Cornufer* inhabits New Guinea, New Britain, the Admiralty Islands, Bougainville, and the Solomon Islands, with outlying species in Seram, Indonesia, the Palau Islands, and Fiji.

Many of the frogs in these two genera are rough-skinned, drably colored terrestrial species, such as the Papuan Wrinkled Ground Frog (*C. papuensis*) and the Philippine Wrinkled Ground Frog (*P. corrugatus*). The large Günther's Wrinkled Ground Frog (*C. guentheri*) is a terrestrial sit-and-wait ambusher which preys on a range of invertebrates but also takes small reptiles and frogs, including its own species. Some species are especially aquatic and possess extensive toe webbing, for example

DISTRIBUTION
Philippines, Indonesia, Papua New Guinea, Palau, Solomon Islands, and Fiji

GENERA
Cornufer, Platymantis

HABITATS
Rainforest, secondary forest, plantations, gardens, and islands

SIZE
½ in (15 mm) ♀ Sierra Madre Pygmy Wrinkled Ground Frog (*P. pygmaeus*) to 9¾ in (250 mm) ♀ Guppy's Giant Water Frog (*C. guppyi*)

ACTIVITY
Nocturnal; terrestrial, arboreal, aquatic, saxicolous, cavernicolous

REPRODUCTION
Direct breeding, known species laying small numbers of large eggs

DIET
Insects, arachnids, also small reptiles and frogs (*C. guentheri*)

IUCN STATUS
CR = 1, EN = 9, VU = 10, NT = 11; percentage species in trouble = 34%

the largest species, Guppy's Giant Water Frog (*C. guppyi*), and the toad-like Warty Webbed Frog (*C. bufoniformis*). Still other species are adapted to live in limestone karst habitats where they enter caves, such as the Critically Endangered South Gigante Cave Frog (*P. insulatus*) and the Negros Island Cave Frog (*P. spelaeus*).

Despite being generally considered terrestrial frogs, the genera *Cornufer* and *Platymantis* also contain a number of arboreal species with expanded terminal discs on their toes and smooth dorsal skin, such as the banana-dwelling Wolf's Sticky-toed Frog (*C. wolfi*), the Solomon Islands Palmate Frog (*C. heffernani*), and Hazel's Forest Frog (*P. hazelae*). Only two native frogs inhabit Fiji, the Fijian Ground Frog (*C. vitianus*) and the Fijian Treefrog (*C. vitiensis*).

Apart from the South Gigante Cave Frog, a further nine ceratobatrachine species are listed as Endangered and ten are Vulnerable.

INDIAN DANCING FROGS

Micrixalidae contains a single genus, *Micrixalus* (24 spp.), the dancing frogs, which was previously included in the Ranidae. Fourteen of the known species were described using molecular techniques in 2014. *Micrixalus* is endemic to the Western Ghats of southwestern India, from southern Maharashtra through Karnataka and Kerala to Tamil Nadu. They demonstrate again the importance of these ancient, rainforested mountain ranges as a biodiversity hotspot (see also Nyctibatrachidae, page 204).

Dancing frogs are relatively small with pointed heads and a diverse array of dorsal patterns. They are found along fast-flowing hill streams, where their calls may be drowned out by the sound of the water. Frogs occurring in this sort of habitat are usually referred to as "torrent frogs." One species, the Cave Dancing Frog (*M. spelunca*), inhabits caves.

The most-studied species is the Waynad Dancing Frog (*M. saxicola*). Male dancing frogs are territorial, jealously guarding their display rocks against rival males. Their tactics involve a series of choreographed leg-lengthening and toe-splaying exercises known as "foot-flagging." Some species even resort to kicking rival males. They use the same tactics in combination with a call to attract females, and it is this behavior that earns them the moniker of "dancing frogs." A pair will then engage in amplexus in the water and the eggs are either laid on rocks along the edge of the stream, where they will keep moist, or they are laid in small burrows that are then covered over.

The Kottigehar Dancing Frog (*M. kottigerharensis*), which has an immaculate white throat-pouch that appears as a flash of white when the male displays, is Critically Endangered, Gadgil's Dancing Frog (*M. gadgili*)

DISTRIBUTION
Southwestern India

GENUS
Micrixalus

HABITATS
Tropical rainforest, evergreen forest, and montane forest alongside forest streams

SIZE
½ in (13 mm) Elegant Dancing Frog (*M. elegans*) to 1¼ in (33 mm) Dusky Dancing Frog (*M. fuscus*) or Kottigehar Dancing Frog (*M. kottigerharensis*)

ACTIVITY
Diurnal; terrestrial, aquatic

REPRODUCTION
Mating is in the water; eggs are laid along the stream edge or in small chambers

DIET
Small invertebrates

IUCN STATUS
CR = 1, EN = 1, VU = 3, NT = 1; percentage species in trouble = 25%

is Endangered, a further three species are Vulnerable, and one is Near Threatened. Other species, like the Dusky Dancing Frog (*M. fuscus*), may be common within their range, but the concern is that most of these frogs have very small ranges and only 3 occur in protected areas, 14 in partially protected areas, and 7 in unprotected areas.

OPPOSITE | The Kottigehar Dancing Frog (*Micrixalus kottigeharensis*) is Critically Endangered.

ABOVE | A pair of territorial male Kottigehar Dancing Frogs (*Micrixalus kottigeharensis*) "foot-flagging."

RIGHT | The Northern Dancing Frog (*Micrixalus uttaraghati*) sheltering in a water-filled bracket fungus.

INDIAN LEAPING FROGS

LEFT | The Critically Endangered Warty-skinned Leaping Frog (*Walkerana phrynoderma*) is listed as one of the 100 EDGE (Evolutionarily Distinct and Globally Endangered) amphibians.

OPPOSITE | Beddome's Leaping Frog (*Indirana beddomii*) is a commonly encountered frog in southern India.

OPPOSITE BELOW | Gundia Frogs (*Indirana gundia*) are also listed as Critically Endangered.

Ranixalidae contains two genera, *Indirana* (14 spp.) and *Walkerana* (4 spp.). Members of both genera are known as leaping frogs. *Indirana* species are generally slightly larger than *Walkerana* species, but all members of the family are fairly small frogs and resemble ranids in appearance, being fairly robust with powerful hindlimbs, pointed heads, large eyes, and often a dark postocular stripe obscuring the tympanum. Toe webbing is more complete in *Indirana*.

Indian leaping frogs are endemic to the Western Ghats of southwestern India, from Gujarat to Tamil Nadu. East–west mountain passes through the Western Ghats represent important divisions between species. One such is the Palakkad or Palghat Gap separating the Anaimalai Hills in the south from the Nilgiri Hills to the north. Beddome's Leaping Frog (*I. beddomii*), the Bhadrai Leaping Frog (*I. bhadrai*), the Rocky Terrain Leaping Frog (*I. paramakri*), the Netravali Leaping Frog (*I. salelkari*),

DISTRIBUTION
Southwestern India (Western Ghats)

GENERA
Indirana, Walkerana

HABITATS
Deciduous and evergreen forest, stream banks, paddy fields, and rocky outcrops

SIZE
1 in (27 mm) ♀ Yadera Leaping Frog (*I. yadera*) to 2 in (54 mm) ♀ Gundia Frog (*I. gundia*)

ACTIVITY
Poorly known; diurnal, nocturnal; terrestrial

REPRODUCTION
Poorly known; tadpoles are semiterrestrial on rocks

DIET
Small invertebrates

IUCN STATUS
CR = 2, EN = 3, VU = 1, percentage species in trouble = 33%

and the Muduga Mountain Leaping Frog (*W. muduga*) are found to the north, while Günther's Leaping Frog (*I. brachytarsus*), Sarojamma's Leaping Frog (*I. sarojamma*), the Yadera Leaping Frog (*I. yadera*), and Spotted Leaping Frog (*W. diplosticta*) occur to the south. The Half-webbed Spotted Frog (*I. semipalmata*) occurs on both sides of the Gap, suggesting it may comprise two morphologically similar species. Further north, the Goa Gap also acts as a similar barrier.

Indian leaping frogs inhabit deciduous and evergreen forests up to elevations of 5,250 ft (1,600 m) ASL, rocky outcrops, the banks of small streams, and caves. They may be encountered both day and night, in forest floor leaf-litter.

This is a very poorly studied family and little is known about its natural history, but one unusual trait has been reported. The tadpoles of some species are semiterrestrial and even before they develop limbs they are able to move across the ground powered by their long tails.

The Warty-skinned Leaping Frog (*W. phrynoderma*) and the Gundia Frog (*I. gundia*) are Critically Endangered, three other species are Endangered, and one is Vulnerable.

INDIAN & SRI LANKAN ROBUST FROGS

OPPOSITE | Humayun's Wrinkled Frog (*Nyctibatrachus humayuni*) belongs to a large genus of southern Indian frogs.

BELOW | The only species in the Astrobatrachinae, the Starry Dwarf Frog (*Astrobatrachus kurichiyana*).

The Nyctibatrachidae is an ancient family with three subfamilies. The Astrobatrachinae was only erected in 2019 for the Starry Dwarf Frog (*Astrobatrachus kurichiyana*), from the Kurichiyarmala District of Kerala in the Western Ghats, southern India. It is a small, light-brown and dark-gray frog with a profusion of small white spots on the flanks and undersides and orange under the limbs. Nocturnal and terrestrial in habit, this is a secretive frog that inhabits montane forest leaf-litter, usually near water. Little is known of its natural history.

Lankanectinae is endemic to southern Sri Lanka, and comprises two species, the widespread Sri Lankan Corrugated Frog (*Lankanectes corrugatus*) and the recently described Knuckles Range Corrugated Frog (*L. pera*), named for the University of Peradeniya, Kandy. Both species are fully aquatic, flattened frogs with a series of ridges across their backs. *Lankanectes corrugatus* occurs between 195–5,000 ft (60–1,525 m) elevation, *L. pera* occurs around 3,610 ft (1,100 m) elevation. Corrugated frogs have fang-like teeth in their mouths, to assist them when feeding on aquatic invertebrates, but they also take prey as diverse as dragonflies, centipedes, and

DISTRIBUTION
Southern India and Sri Lanka

GENERA
(1) *Astrobatrachus*; (2) *Lankanectes*; (3) *Nyctibatrachus*

HABITATS
Montane forest, evergreen forest, forest torrents, marshes, and rice paddies

SIZE
(1) 1 in (27 mm) ♂ to 1¼ in (29 mm) ♀ Starry Dwarf Frog (*A. kurichiyana*); (2) ⁵⁄₁₆ in (8 mm) Knuckles Range Water Frog (*L. pera*) to 2¾ in (71 mm) ♀ Corrugated Water Frog (*L. corrugatus*); (3) ½ in (13 mm) Kakkachi Night Frog (*N. beddomii*) to 3¼ in (84 mm) Giant Wrinkled Frog (*N. karnatakaensis*)

ACTIVITY
Nocturnal, occasionally diurnal; terrestrial, aquatic

REPRODUCTION
Largely unknown, no amplexus in some species (*N. humayuni*), eggs laid in vegetation overhanging ponds, tadpoles dropping into the water

Key: (1) Southern India (red), (2) Sri Lanka (blue)

DIET
Presumed small invertebrates

IUCN STATUS
CR = 3, EN = 5, VU = 3;
NT = 1; percentage species
in trouble = 32%

earthworms. Nocturnal, their reproductive habitats are unstudied. *Lankanectes corrugatus* likes muddy-bottomed watercourses but is Critically Endangered because of pollution, while the sandy-bottom dwelling *L. pera* is Near Threatened.

Nyctibatrachinae also contains a single genus, *Nyctibatrachus* (36 spp.), of night frogs and wrinkled frogs, found in the Western Ghats from Maharashtra to Tamil Nadu, but with most species at the southern end of the range. Generally associated with rocky streams in montane evergreen forests, some species inhabit marshes. They are nocturnal, terrestrial, and arboreal. The Humayun's Wrinkled Frog (*N. humayuni*) engages in a strange ritual where the female backs into the male, which engages in amplexus but then dismounts from the female before she deposits her eggs on leaves overhanging water into which the hatching tadpoles will drop.

Dattatreya's Night Frog (*N. dattatreyaensis*) and the Coorg Night Frog (*N. sanctipalustris*) are Critically Endangered, five species are Endangered, and three are Vulnerable.

ABOVE LEFT | The Sri Lankan Corrugated Frog (*Lankanectes corrugatus*) is one of only two species in the Sri Lankan endemic Lankanectinae.

FORK-TONGUED FROGS

OPPOSITE | A salt-water tolerant anuran, the Crab-eating Frog (*Fejervarya crancrivora*) lives in mangrove swamps and feeds on marine crustaceans.

BELOW | The Indian Bullfrog (*Hoplobatrachus tigrinus*), at equilibrium in India, has been introduced to Madagascar and may threaten native frogs.

The Dicroglossidae comprises 2 subfamilies, 15 genera, and 214 species of primarily Asian frogs. Dicroglossinae occurs from Iran to Japan and south to New Guinea, with the Arabian Skittering Frog (*Euphlyctis ehrenbergii*) in southwest Arabia, and the African Bullfrog (*Hoplobatrachus occipitalis*) the only African species. Widely distributed across Sub-Saharan Africa, its tadpoles eat malarial mosquito larvae. The Indian Bullfrog (*H. tigerinus*) is introduced to Madagascar, and may threaten native anurans. The Indo-Malaysian Crab-eating Frog (*Fejervarya cancrivora*) is the most salt-tolerant anuran, inhabiting mangrove swamps. Other *Fejervarya* live in rice paddies, for example, the Lesser Sunda Rice-paddy Frog (*F. verruculosa*).

Limnonectes (77 spp.) is the largest genus, and includes the largest species, Blyth's Fanged Frog (*L. blythi*) from Southeast Asia. Fanged frogs have enlarged teeth on their mandibles for use in combat or hunting. The Giant Borneo River Frog (*L. leporinus*) is large enough to eat other frogs. Usually females are larger than males but the male Khorat Fanged Frog (*L. megastomias*) reverses this trend, and reputedly occasionally eats birds. Not all dicroglossines are carnivores; the Indian Bullfrog (*Phrynoderma hexadactylum*) also eats vegetation.

DICROGLOSSINAE

DISTRIBUTION
Asia, New Guinea, and Africa

GENERA
Allopaa, Chrysopaa, Euphlyctis, Fejervarya, Hoplobatrachus, Limnonectes, Minervarya, Nannophrys, Nanorana, Ombrana, Phrynoderma, Quasipaa

HABITATS
Most habitats, forest, swamps, grassland, ponds, mangroves (*F. cancrivora*), and mountains

SIZE
¾ in (19 mm) Southern Cricket Frog (*M. syhadrensis*) to 10¼ in (260 mm) Blyth's Fanged Frog (*L. blythi*)

ACTIVITY
Nocturnal; terrestrial, aquatic

REPRODUCTION
Axillary amplexus; aquatic larvae or direct breeding

DIET
Invertebrates including crabs, other frogs, and vegetation

IUCN STATUS
EX = 1, CR = 1, EN = 14, VU = 16, NT = 13; percentage species in trouble = 23%

Nanorana (31 spp.) occurs high in the Himalayas, for example, the Xizang Plateau Frog (*Nanorana parkeri*), between 9,350–16,400 ft (2,850–5,000 m). The monotypic Sikkim Frog (*Ombrana sikimensis*) is also Himalayan. Many of the smaller dicroglossine frogs are aquatic, like the Common Skittering Frog (*Euphlyctis cyanophlyctis*), which adopts a splayed-legged posture on the water surface, skittering away at any sign of danger. Other species are more rounded and robust, for example, the Indian Burrowing Frog (*Sphaerotheca breviceps*).

Occidozygyginae contains two genera, the largest being *Occidozyga* (16 spp.), the oriental frogs. Occidozygine frogs are aquatic and inhabit flooded grasslands, swamps, and streams.

Günther's Streamlined Frog (*Nannophrys guentheri*), from Sri Lanka, is believed Extinct, while the Charles Darwin's Frog (*Minervarya charlesdarwini*), from the Andaman Islands, and the Mount Tompotika Frog (*Occidozyga tompotika*), from Sulawesi, Indonesia, are Critically Endangered. A further 14 species are Endangered, 18 Vulnerable, and 14 Near Threatened.

RIGHT | The Philippine Puddle Frog (*Occidozyga laevis*) has dorsally positioned eyes so it does not have to expose its entire head when at the surface.

OCCIDOZYGINAE

DISTRIBUTION
Asia

GENERA
Ingerana, Occidozyga

HABITATS
Flooded grassland, swamps, stagnant ponds, and streams

SIZE
¾ in (18 mm) Reticulate Trickle Frog (*I. reticulata*) to 2½ in (60 mm) Philippine Oriental Frog (*O. laevis*)

ACTIVITY
Nocturnal; semiaquatic, aquatic

REPRODUCTION
Usually axillary amplexus; aquatic larvae or direct breeding

DIET
Aquatic and terrestrial invertebrates

IUCN STATUS
CR = 1, VU = 2, NT = 1; percentage species in trouble = 21%

WEST AFRICAN TOOTHED FROGS

Odontobatrachidae is one of the smallest African frog families, comprising a single genus, *Odontobatrachus* (5 spp.). Until 2015 only a single species was recognized, the Common Tooth Frog (*O. natator*), described in 1905, but an in-depth molecular study revealed four additional cryptic species.

Odontobatrachus tooth frogs are endemic to the rainforests of West Africa. The Common Tooth Frog occurs in southeastern Guinea, Sierra Leone, and Liberia, while the newly described species are Arndt's Tooth Frog (*O. arndti*) from Guinea, Liberia, and Côte d'Ivoire; the Fouta Djallon Tooth Frog (*O. fouta*) and Smith's Tooth Frog (*O. smithi*) from

DISTRIBUTION	SIZE	DIET
West Africa	2 in (50 mm) ♂ Ziama Toothed Frog (*O. ziama*) to 2½ in (64 mm) ♀ Arndt's Toothed Frog (*O. arndti*)	Diverse, includes vegetation and other frogs

GENUS
Odontobatrachus

HABITATS
Fast-flowing rainforest streams and cascades to 4,265 ft (1,300 m)

ACTIVITY
Nocturnal; terrestrial, aquatic, occasionally arboreal

REPRODUCTION
Eggs laid on land, tadpoles adapted for torrent living

IUCN STATUS
Data not available

western Guinea; and the Ziama Tooth Frog (*O. ziama*) from southeastern Guinea. This last species has been recorded up to 4,265 ft (1,300 m) elevation. All five species inhabit fast-flowing streams and rivers, in particular the rocks and vegetation around rocky torrents. Perhaps further species remain to be discovered in the West African rainforests.

West African tooth frogs are moderately large frogs with rugose granular skin and glandular ridges, and fingers and toes with bifurcated terminal discs. Males, rarely females, also exhibit large oval glands on the undersides of their femurs, the purpose of which is unknown. The most distinctive feature of these frogs is to be found inside their mouths.

Both sexes possess enlarged backward-facing teeth on the maxillary and premaxillary bones and a pair of tusklike fangs in the mandibles. Such dentition may be used to enable the tooth frogs to overpower and swallow other frogs. Their alternative name is saber-toothed frogs. These fangs are also used in fights between rival males and could cause considerable injury. Surprisingly tooth frogs are also thought to include vegetation in their diets. Tooth frogs lay their eggs on land in the splash zone, and their tadpoles are streamlined with large oral suckers to enable them to anchor to rocks and swim against the current.

ABOVE | Arndt's Tooth Frog (*Odontobatrachus arndti*) inhabits the West African Nimba Mountains and was identified by molecular data.

OPPOSITE | From 1905 until 2015 the Common Tooth Frog (*Odontobatrachus natator*) was thought to be a single species.

SHOVEL-NOSED FROGS

Hemisotidae is a small sub-Saharan African family most closely related to the Brevicipitidae, the rain frogs. It comprises the genus *Hemisus* (9 spp.), the shovel-nosed frogs, also known as piglet frogs. These are rotund frogs with short, sharply pointed snouts, small eyes, and short legs. They are forward burrowers that use their strengthened snouts to excavate burrows, pushing forward with their powerful hindlimbs and flicking soil backward with the forelimbs. They occur in a wide range of forest and savanna habitats.

LEFT | The Marbled Shovel-nosed Frog (*Hemisus marmoratus*) is the most widely distributed species across West, Central, and East Africa.

ABOVE | De Witte's Shovel-nosed Frog (*H. wittei*) inhabits the Upemba National Park in southern DRC and northwestern Zambia.

OPPOSITE | Perret's Shovel-nosed Frog (*H. perreti*) inhabits rainforest on the Atlantic coast from Gabon to the DRC.

DISTRIBUTION
Sub-Saharan Africa

GENUS
Hemisus

HABITATS
Lowland and highland tropical rainforest, tropical and subtropical savanna, savanna woodland, and gallery forest

SIZE
1¼ in (30 mm) ♂ Barotse Shovel-nosed Frog (*H. barotseensis*) to 3 in (80 mm) ♀ Spotted Shovel-nosed Frog (*H. guttatus*)

ACTIVITY
Nocturnal; fossorial

REPRODUCTION
Inguinal amplexus; the female builds an incubation chamber and cares for eggs and larvae

DIET
Nocturnal termites, also earthworms

IUCN STATUS
NT = 1; percentage species in trouble = 11%

Two species are widely distributed across Central and West Africa, the Guinea Shovel-nosed Frog (*H. guineensis*) and the Marbled Shovel-nosed Frog (*H. marmoratus*), but the other species are more localized. The Short-fingered Shovel-nosed Frog (*H. brachydactylus*), is found in central Tanzania; the Barotse Shovel-nosed Frog (*H. barotseensis*), in northwest Zambia; both De Witte's Shovel-nosed Frog (*H. wittei*) and the Olive Shovel-nosed Frog (*H. olivaceus*), in the Democratic Republic of the Congo; Perret's Shovel-nosed Frog (*H. perreti*), along the coast from Gabon to the Democratic Republic of the Congo, and the Spotted Shovel-nosed Frog (*H. guttatus*), the southernmost species, in KwaZulu-Natal. Most species occur in lowland habitats but the Ethiopian Shovel-nosed Frog (*H. microscaphus*) occurs in the Ethiopian Highlands at 4,920–8,860 ft (1,500–2,700 m) elevation.

Reproduction takes place at the start of the rains when the male calls to attract a female. The female then lays between 200 (*H. marmoratus*) and 2,000 (*H. guttatus*) eggs in an incubation burrow she has constructed, remaining with them until they hatch, whereupon she will breach the chamber to allow the tadpoles to reach the water. Because they utilize temporary pools the free-swimming tadpoles must develop into froglets in under a month. Shovel-nosed frogs feed on nocturnal termites, but will also eat earthworms.

Only one species, the Spotted Shovel-nosed Frog, is considered Near Threatened, but the Short-fingered, Barotse, and De Witte's shovel-nosed frogs are listed as Data Deficient, which means we do not know their true conservation status. The other five species are listed as Low Concern.

AFRICAN RAIN FROGS

LEFT | The Bushveld Rain Frog *(Breviceps adspersus)* is the most widely distributed member of the genus in Southern Africa.

OPPOSITE | An endemic from the Ethiopian Bale Mountains, the Bale Rain Frog *(Balebreviceps hillmani)* is Critically Endangered.

OPPOSITE BELOW | The Mazumbai Warty Frog *(Callulina kisiwamsitu),* from the Tanzanian Usambara Mountains, is Endangered.

Brevicipitidae contains 5 genera and 37 species across sub-Saharan Africa. *Breviceps* (20 spp.) is the largest and most widely distributed genus. These secretive frogs spend much of their time in self-dug burrows, excavated by burrowing backward at speed, only emerging after rain. Rain frogs are rotund—in some cases, for example, the Strawberry Rain Frog *(Br. acutirostris),* excessively so—it resembles a small inflated balloon with four legs and a face, but with their down-turned mouths, all rain frogs present a "not happy" impression.

They inhabit semiarid savanna and desert grasslands from Tanzania to South Africa, the Bushveld Rain Frog *(Br. adspersus)* occupying much of the generic range. South Africa has 17 species, 13 being endemic. The name of one species aptly reflects their burrowing habits: Bilbo Baggins' Rain Frog *(Br. bagginsi).*

At elevations of 9,840 ft (3,000 m) ASL in the Ethiopian Bale Mountains occurs the Critically Endangered Bale Rain Frog *(Balebreviceps hillmani).* Males do not call to attract a mate, are nocturnal,

DISTRIBUTION
East Africa

GENERA
Balebreviceps, Breviceps, Callulina, Probreviceps, Spelaeophryne

HABITATS
Arid and semiarid desert, grassland, savanna woodland, and montane forest

SIZE
½ in (15 mm) ♂ Rose's Rain Frog *(Br. rosei)* to 3 in (80 mm) ♀ Giant Bush Frog *(Br. gibbosus)*

ACTIVITY
Nocturnal; terrestrial, fossorial, arboreal

REPRODUCTION
Direct breeders; eggs laid in burrows hatch into froglets

DIET
Terrestrial invertebrates from ants to beetles

IUCN STATUS
CR = 8, EN = 7, NT = 4; percentage species in trouble = 51%

and spend the day under stones. Genus *Probreviceps* (6 spp.) occurs in the forests of East Africa's Rift Mountains. Five of the species are endemic to Tanzania and one, the Highland Forest Frog (*P. rhodesianus*), to Zimbabwe, but all six are Endangered.

Genus *Callulina* (9 spp.) is also primarily Tanzanian, with just the Taita Warty Frog (*C. dawida*), inhabiting Kenya. Warty frogs are small, semiarboreal, nocturnal, montane forest dwellers. Seven species are Critically Endangered, and one Endangered. The final genus is monotypic, containing Methner's Cave Frog (*Spelaeophryne methneri*), a rarely encountered Tanzanian species which is black with red chevron markings.

The short limbs of the males make gripping the larger females difficult, so they glue themselves to the female's back during amplexus. They are direct breeders; eggs laid in burrows hatch directly into froglets. The Large Forest Frog (*P. macrodactylus*) guards its nest during incubation. Defensive responses include body inflation, secretion of noxious chemicals, and in the case of *S. methneri*, dermal autotomy, shedding its own skin.

ABOVE | Vibeke's Spiny Reed Frog (*Afrixalus vibekensis*) is an attractive species from Ivory Coast and Ghana.

Hyperoliinae contains nine genera, including *Hyperolius* (143 spp.), the reed frogs, Africa's largest anuran genus. Reed frogs are small, smooth-skinned, with discs on their webbed digits, horizontal pupils, and hidden tympana, that perch on reeds around ponds. Many are polymorphic, for example, the Variable Reed Frog (*Hy. pictus*), while others are confusingly similar—pale green with pale-yellow stripes a common theme. Some change color from day to night, still others are sexually dichromatic. They occupy many habitats, including at high elevations, for example, the Brown Reed Frog (*Hy. castaneus*) at 5,250–9,350 ft (1,600–2,850 m), in the Democratic Republic of the Congo (DRC).

Although present in every sub-Saharan country, the DRC (45 spp.), Tanzania (39 spp.), and Cameroon (31 spp.), demonstrate the greatest diversity. Bioko Island has three species, São Tomé and Príncipe has three endemics, while Unguja (Zanzibar) and Pemba host two endemic species. Some females guard their eggs, aid hatching, and guide the larvae to water. The Rumpi Hills Egg-guarding Frog (*Hy. jynx*) is Critically Endangered.

Afrixalus (35 spp.), the spiny reed frogs, have spinous heads and bodies, and vertically elliptical pupils. Diversity is greatest in the DRC (12 spp.), Tanzania (11 spp.), and Cameroon (11 spp.).

DISTRIBUTION
Sub-Saharan Africa, Bioko Island, São Tomé and Príncipe, Unguja and Pemba Islands, Madagascar, and the Seychelles

GENERA
Afrixalus, Congolius, Cryptothylax, Heterixalus, Hyperolius, Kassinula, Morerella, Opisthothylax, Tachycnemis

HABITATS
Primary rainforest, degraded forest, gallery forest, dry forest, coastal forest, savanna woodland, flooded grassland, agricultural land, swamps, rivers, streams, ponds, arid grassland, lowland, highland, and islands

SIZE
¾ in (17 mm) ♂ Dwarf Reed Frog (*Hy. minutissimus*) to 3 in (76 mm) ♀ Seychelles Treefrog (*T. seychellensis*).

ACTIVITY
Nocturnal; arboreal, scansorial in riparian or lacustrine vegetation

The Ethiopian Highlands contain two endemics, Clarke's Spiny Reed Frog (*A. clarkei*, 2,690–6,660 ft [820–2,030 m]) and the Grassland Spiny Reed Frog (*A. enseticola*, 5,580–9,020 ft [1,700–2,750 m]), one Endangered, the other Vulnerable. Fornasini's Spiny Reed Frog (*A. fornasini*) eats the eggs of other frogs.

The Malagasy reed frogs, *Heterixalus* (11 spp.), exhibit a "vertical-rhomboid" pupil—anterior edge angular, posterior edge curved. Highly variable in pattern, calls are required to distinguish species. Unique hyperoliines include the Congo Frog (*Congolius robustus*), two Green-eyed frogs (*Cryptothylax* spp.), a Blue-eyed Frog (*Morerella cyanophthalma*), an Orange Frog (*Opisthothylax immaculatus*), and De Witte's Clicking Frog (*Kassinula wittei*). The Seychelles Treefrog (*Tachycnemis seychellensis*), the largest species, inhabits the Seychelles granitic islands. Five species are Critically Endangered, 16 Endangered, 16 Vulnerable, and 6 Near Threatened.

REPRODUCTION
Axillary amplexus; eggs glued to overhanging leaves or on submerged or floating vegetation; tadpoles carnivorous, except some terrestrial *Hyperolius* (egg guarders)

DIET
Probably small invertebrates and other frogs' eggs (*A. fornasini*)

IUCN STATUS
CR = 5, EN = 18, VU = 16, NT = 6; percentage species in trouble = 23%

ABOVE | Common Reed Frogs (*Hyperolius viridiflavus*) appear in at least 50 different color patterns.

INSET | *Heterixalus* is endemic to Madagascar. Pictured is the Malagasy Reed Frog (*H. madagascariensis*) from the east of the country.

AFRICAN RUNNING FROGS, WOT-WOTS & SPINY FROGS

Kassininae contains four genera. *Kassina* (15 spp.) are the African running frogs, with the Senegal Running Frog (*K. senegalensis*) widely distributed from Senegal to Somalia, to South Africa, where it overlaps the range of the monotypic Weale's Running Frog (*Semnodactylus wealii*). The Jozani Running Frog (*K. jozani*) is endemic to Unguja (Zanzibar). These are boldly patterned frogs with rows of large, dark blotches on a pale background. Primarily terrestrial, they are excellent climbers, inhabiting forests and savannas, but spending much of their time hidden in burrows, only emerging after rain.

Paracassina (2 spp.) are the Ethiopian mountain running frogs. The Kouni Valley Running Frog (*P. kounhiensis*) inhabits marshes in the Rift Mountains at 6,495–10,500 ft (1,980–3,200 m) elevation, where it feeds on mollusks, and is considered Vulnerable.

The forest-dwelling wot-wots, *Hylambates* (5 spp.), named for their distinctive calls, "wot, wot," are larger than running frogs, and exhibit a scattered distribution from Sierra Leone to South Africa. The Eastern Wot-Wot (*H. maculatus*) has bright red flanks, while Keith's Wot-Wot (*H. keithae*), from Tanzania, has yellow flanks. When threatened they contort their bodies to expose these colors, similar to the "unkenreflex" posture of fire-bellied toads (*Bombina*, see page 66).

DISTRIBUTION
Sub-Saharan Africa, Bioko Island, and Unguja Island

GENERA
(1) *Hylambates, Kassina, Paracassina, Semnodactylus*;
(2) *Acanthixalus*;
(3) *Arlequinus, Callixalus, Chrysobatrachus*

HABITATS
Rainforest, edge situations, wet or arid savana, thornbush scrub, montane forest or grassland, swamps, fynbos, gardens, and bamboo thickets

SIZE
(1) 1¼ in (32 mm) Decorated Running Frog (*K. decorata*) to 2¾ in (72 mm) Eastern Wot-Wot (*H. maculatus*);
(2) 1½ in (39 mm) Sonja's Spiny Frog (*A. sonjae*) or Central African Spiny Frog (*A. spinosus*); (3) 1 in (24 mm) ♂ Itombwe Copper Frog (*Ch. cupreonitens*) to 1¾ in (43 mm) ♀ Congo Painted Frog (*Ca. pictus*)

Key: (1) Kassinae (red, blue & green), (2) Acanthixalinae (blue), (3) *Incertae sedis* (green)

Acanthixalinae contains just the genus *Acanthixalus* (2 spp.). The Central African Spiny Frog (*A. spinosus*) is an excessively spiny frog with large eyes and pupils that contract to a diamond shape. The Near Threatened Sonja's Spiny Frog (*A. sonjae*), from West Africa, is less rugose. Spiny frogs are aquatic and semiarboreal, using water-filled tree holes for egg-laying. When threatened they play dead (thanatosis), flattening their bodies, closing their eyes, and protruding their tongues.

Three monotypic genera are *incertae sedis*. The Harlequin Frog (*Arlequinus krebsi*) is a small, Endangered treefrog from Cameroon and Bioko Island. The Itombwe Copper Frog (*Chrysobatrachus cupreonitens*), patterned orange and green with black speckles, from the Itombwe Mountains, Democratic Republic of the Congo, is also Endangered, while the pale-yellow Painted Frog (*Callixalus pictus*), also from the Itombwe range, is Vulnerable.

ACTIVITY
Nocturnal, esp. after rain, or diurnal; arboreal, terrestrial, aquatic

REPRODUCTION
Axillary amplexus, (1) breed in pools or tree holes, 1–600 eggs; (2) 5–13 eggs in tree holes or branches over water, tadpoles carnivorous; (3) little-known, < 16 eggs in still or slow water (*Arlequinus*)

DIET
Probably small invertebrates, slugs, and snails (*Paracassina*)

IUCN STATUS
EN = 4, VU = 4, NT = 3; percentage species in trouble = 39%

TOP | From montane swamps and seasonally flooded grasslands, the Itombwe Copper Frog (*Chrysobatrachus cupreonitens*) is an Endangered species.

ABOVE | The Central African Spiny Frog (*Acanthixalus spinosus)* is found from southern Nigeria to eastern DRC.

SQUEAKERS & LONG-FINGERED FROGS

Arthroleptidae contains three subfamilies, the Arthroleptinae being found in tropical Africa and comprising two genera: *Arthroleptis* (48 spp.), the squeakers, so-called because of their high-pitched insect-like calls; and *Cardioglossa*, 19 spp. of long-fingered frogs, in which the third finger of males is exceedingly long and bears small spines on its ventral surface. Some species of *Arthroleptis* also develop elongate third fingers. A few species, for example, the Plain Squeaker (*A. xenochirus*), have an exceedingly long first finger which looks remarkably ungainly.

These are all small, leaf-litter inhabitants, with most *Arthroleptis* cryptically patterned to avoid detection, but a few, like the Beautiful Squeaker (*A. formosus*) from Guinea, are more boldly patterned with red or yellow markings. Many *Cardioglossa* have bold black marbling along the flanks and blues and reds in their patterning.

DISTRIBUTION
Tropical Africa

GENERA
Arthroleptis, Cardioglossa

HABITATS
Rainforest, montane forest, savanna, and savanna woodland streams to 8,860 ft (2,700 m)

SIZE
½ in (15 mm) ♂ Vercammen's Squeaker (*A. vercammeni*) or ♀ Hidden Squeaker (*A. fichika*) to 2¼ in (58 mm) ♀ Nguru Squeaker (*A. nguruensis*)

ACTIVITY
Diurnal, nocturnal; terrestrial, aquatic

REPRODUCTION
Axillary amplexus; eggs laid in streams with tadpole stage (*Cardioglossa*), or leaf-litter or burrows for direct development (*Arthroleptis*)

DIET
Small invertebrates

IUCN STATUS
CR = 6, EN = 11, VU = 4, NT = 3; percentage species in trouble = 36%

The greatest diversity of *Arthroleptis* is found in Tanzania (15 spp.), the Democratic Republic of the Congo (DRC, 13 spp.), Cameroon (9 spp.) and Gabon (6 spp.). Three species occur on Bioko Island, Gulf of Guinea, including the endemic Bioko Squeaker (*A. bioko*), and the southernmost species, the Bush Squeaker (*A. wahlbergii*), enters KwaZulu-Natal. *Cardioglossa* is primarily Central African, with Cameroon (14 spp.), DRC (6 spp.), Gabon (5 spp.) demonstrating the greatest diversity, but it is absent from Tanzania.

The Black-spotted Long-fingered Frog (*C. nigromaculata*) inhabits Nigeria, Cameroon, and Bioko Island.

Cardioglossa live along streams and lay eggs that hatch into free-swimming tadpoles into the water. *Arthroleptis* often occur some distance from water and are direct developers, using terrestrial nests for eggs that hatch into small froglets. Some species demonstrate ontogenetic changes in diet, for example, froglets of the Dwarf Squeaker (*A. xenodactyloides*), from the Usambara Mountains, Tanzania, feed on springtails, but prey on ants when they become adult.

Six species are Critically Endangered, including the Cave Squeaker (*A. troglodytes*) from Zimbabwe, Nike's Squeaker (*A. nikeae*) from Tanzania, and the Manengouba Long-fingered Frog (*C. manengouba*) from Cameroon. Additionally, 11 species are Endangered, 4 are Vulnerable, and 3 Near Threatened.

OPPOSITE | The Krokosua Squeaker (*Arthroleptis krokosua*) is only found on two mountains in Ghana and Guinea and listed as Near Threatened.

BELOW | The attractive Black-spotted Long-fingered Frog (*Cardioglossa nigromaculata*) occurs in southern Cameroon and on Bioko Island.

NIGHT, HAIRY & EGG FROGS

LEFT | The Gabon Forest Frog (*Scotobleps gabonicus*) has sharp claws on its toes that can deliver a deep wound if it is handled.

OPPOSITE INSET | Cat-eyed Frogs (*Nyctibates corrugatus*) produce elongate tadpoles that resemble eels.

OPPOSITE BELOW | Male Hairy Frog (*Astylosternus robustus*) develop hairlike fringes on their hindlimbs that act like gills and prolong the time the males can stay submerged while egg-guarding.

Astylosterninae contains just four genera. *Astylosternus* (13 spp.) contains the squat-bodied, large-eyed, night frogs, from the rainforests of West and Central Africa. Cameroon alone has 11 species, including Bates' Night Frog (*A. batesi*), females of which may reach 3 in (74 mm) SVL and males 2 in (53 mm).

Included here is the famous Hairy Frog (formerly *Trichobatrachus robustus*), which molecular studies have established belongs in *Astylosternus*, as *A. robustus*. Hairy frogs occur from Cameroon to

Angola. Males are larger than females, the reverse situation to most anurans, and during the breeding season they develop dense fringes of "hair" along their flanks and hindlimbs. The hairs contain blood vessels and are highly vascularized; they act like gills, providing oxygen while the males are submerged guarding their eggs.

Night frogs, the Hairy Frog, and the monotypic Gabon Forest Frog (*Scotobleps gabonicus*), a squat, excessively warty, frog with the same range as the Hairy Frog, are prey to larger vertebrates, including

---

DISTRIBUTION
Tropical West and Central Africa

GENERA
Astylosternus, Leptodactylodon, Nyctibates, Scotobleps

HABITATS
Lowland and montane rainforest streams

SIZE
¾ in (22 mm) ♂ Stevart's Egg Frog (*L. stevarti*) to 5 in (130 mm) ♂ Hairy Frog (*A. robustus*)

ACTIVITY
Diurnal, nocturnal; terrestrial, semi-fossorial

REPRODUCTION
Eggs laid in streams and incubated by male

(*A. robustus*), or in leaf-litter, with tadpole stage

DIET
Invertebrates, e.g., insects, arachnids, myriapods, and mollusks

IUCN STATUS
CR = 3, EN = 12, VU = 3, NT = 4; percentage species in trouble = 73%

humans who especially prize the flesh of the Hairy Frog, but these frogs are not without defenses. They have sharp, keratinized, clawlike phalanges under the skin of their toes, with which to deliver a slashing cut, earning the species the alternative name of Wolverine Frog.. The secretive, nocturnal Cat-eyed Frog (*Nyctibates corrugatus*) is another monotypic species that inhabits rocky streams in the same rainforests as the Gabon Forest Frog and Hairy Frog.

The largest genus *Leptodactylodon* (15 spp.), contains the egg frogs—small, plump, leaf-litter burrowers with rounded snouts and small eyes, that exhibit cryptic dorsal patterns, but which often possess bright-red, white, and black venters. Occurring from Nigeria to Gabon, often several species occur in sympatry, and can only be distinguished by their calls. Egg frog ranges are often very small and vulnerable to habitat change.

Three species are Critically Endangered: the Pale-sided Egg Frog (*L. axillaris*), the Red-bellied Egg Frog (*L. erythrogaster*), and Wild's Egg Frog (*L. wildi*), while 12 are Endangered, 3 Vulnerable, and 4 Near Threatened.

AFRICAN TREEFROGS

The near globally distributed treefrog family Hylidae is conspicuous in its absence from sub-Saharan Africa, while the foam-nesting treefrogs, Rhacophoridae, are represented only by a single small genus (*Chiromantis*, see page 175). The tropical African treefrog niche is therefore occupied by the arthroleptid subfamily Leptopelinae, which contains a single genus, *Leptopelis* (55 spp.). African treefrogs occur throughout tropical Africa from Senegal to Ethiopia, and south to Namibia's Caprivi Strip and KwaZulu-Natal in South Africa.

African treefrogs are large with all the usual attributes of a treefrog, for example, streamlined bodies; long powerful limbs with elongate digits that terminate in large round discs, to provide anchorage when climbing, perching, and landing; large tympana; and large, forward-facing eyes with vertical pupils. They are patterned with greens, browns, and grays to blend into the environment because most species are arboreal in rainforest habitats, a typical species being the Marked Treefrog (*L. notatus*). But some species are more

RIGHT | The Vermiculated Treefrog (*Leptopelis vermicularis*) is an Endangered species from the biodiversity-rich highlands of Tanzania.

DISTRIBUTION
Tropical Africa

GENUS
Leptopelis

HABITATS
Rainforest, savanna woodland, and swamps

SIZE
1 in (26 mm) ♂ Kivu Treefrog (*L. kivuensis*) to

4¼ in (110 mm) ♀ Palm Treefrog (*L. palmatus*)

ACTIVITY
Diurnal, nocturnal; arboreal, terrestrial, semi-fossorial

REPRODUCTION
Axillary amplexus; eggs laid on water or underground nests, with tadpole stage, possible direct development (*L. brevirostris*)

DIET
Small invertebrates, occasionally snails (*L. brevirostris*)

IUCN STATUS
EN = 5, VU = 5, NT = 4; percentage species in trouble = 25%

rotund and terrestrial or semi-fossorial in habit, for example, the Toad-like Treefrog (*L. bufonides*), from the arid Sahel savannas of Senegal to Chad.

The greatest diversity is found in the Democratic Republic of the Congo (18 spp.), Cameroon (14 spp.), Gabon and Tanzania (9 spp. each), and Ethiopia (6 spp.). Outside mainland Africa, the East African Yellow-

spotted Treefrog (*L. flavomaculatus*) also inhabits Unguja (Zanzibar) Island, while Bioko Island has four species, and the largest species, the Palm Treefrog (*L. palmatus*), is endemic to Príncipe Island in São Tomé and Príncipe. The southernmost species are the Mozambique Treefrog (*L. mossambicus*), the Natal Treefrog (*L. natalensis*), and the Long-fingered Treefrog (*L. xenodactylus*), all of which occur in KwaZulu-Natal.

Breeding in the summer rains, females lay eggs in temporary ponds from where tadpoles make their own way to permanent water. The Short-headed Treefrog (*L. brevirostris*) is a direct-breeder, laying eggs in leaf-litter which hatch into froglets. This species is also unusual because it feeds on snails. Five species are Endangered, five Vulnerable, and four Near Threatened.

ABOVE LEFT | Uluguru Treefrogs (*Leptopelis uluguruensis*) also occur in the biologically rich mountains of Tanzania.

ABOVE RIGHT | Mozambique Treefrogs (*Leptopelis mossambicus*) occur as far south as the Lebombo Mountains of KwaZulu Natal.

AMERICAN NARROW-MOUTHED FROGS

LEFT | The Brown Egg Frog (*Ctenophryne geayi*) is a secretive, nocturnal, fossorial species from northern South America.

Microhylids are not small treefrogs, "micro-hylids," quite the opposite, because most of them are terrestrial or fossorial. The Microhylidae is the second-largest anuran family after the Hylidae, with over 400 species and it exhibits an almost global distribution with 12 subfamilies, 3 of which inhabit the Americas.

Most American microhylids belong to the Gastrophryninae, comprising 11 genera, 84 species of small, stout, short-legged, cryptic, fossorial or semi-fossorial species in rainforests, woodlands, and savannas. Males have short legs and cannot grip the female during amplexus, so they glue themselves to the female's back. They choose still water for egg-laying, not being strong swimmers. The largest genus is *Chiasmochelis* (37 spp.), the humming frogs, from Colombia to Argentina, but demonstrating its greatest diversity in the Brazilian Atlantic forests. Genus *Elachistocleis* (22 spp.) contains the oval frogs, although "pear-drop frogs" might be a more appropriate name given their rotund body shape, narrow, pointed heads, and tiny eyes. They occur throughout South America east of the Andes, and also in Panama.

DISTRIBUTION
(1) North, Central & South America; (2,3) northern South America

GENERA
(1) *Arcovomer, Chiasmocleis, Ctenophryne, Dasypops, Dermatonotus, Elachistocleis, Gastrophryne, Hamptophryne, Hypopachus, Myersiella,* *Stereocyclops;* (2) *Otophryne, Synapturanus;* (3) *Adelastes*

HABITATS
(1,2) Rainforest, woodland, savanna, and swamps; (3) Tepui forest

SIZE
(1) ½ in (12 mm) ♂ Ecuador Silent Frog (*Ch. antenori*) to 2¾ in (67 mm) ♀ Costa Rican Egg Frog (*Ct. aterrima*); (2) ¾ in (18 mm) Tapir Frog (*S. danta*) to 2½ in (61 mm) ♀ Pyburn's Pancake Frog (*O. pyburni*); (3) 1¼ in (29 mm) Neblina Forest Frog (*Ad. hylonomus*)

ACTIVITY
Nocturnal; terrestrial, fossorial, secretive

Key: Gastrophryninae (red & blue), (2) Otophryninae (blue), (3) Adelastinae (yellow)

The northernmost gastrophyne genera are *Gastrophryne* (4 spp.) and *Hypopachus* (5 spp.), with three of the former and one of the latter in the United States. The Eastern Narrow-mouthed Toad (*G. carolinensis*) occurs throughout southeast USA.

The other two families are much smaller. Adelastinae contains a single species, the Neblina Forest Frog (*Adelastes hylonomus*), which is only known from three tepui locations in Venezuela, Brazil, and Guyana. Otophryninae contains two genera: *Otophryne* (3 spp.), the pancake frogs, and *Synapturanus* (7 spp.), the disc frogs. Pyburn's Pancake Frog (*O. pyburni*) is a small, dorsally flattened, triangular frog with an extremely pointed snout and a distinctive dorsolateral ridge along its body. The Tapir Frog (*S. danta*) is a rounded frog with a pointed head and a short proboscis-like snout that provides its common name.

Only five American microhylids are listed as Endangered, but given the localized distributions of many species and our lack of knowledge about them, it is likely many more are threatened.

REPRODUCTION
(1) Males glued to females in amplexus; eggs laid in standing water or swamps, or (2) leaf-litter with endotrophic tadpoles; (3) Unknown

DIET
Forest floor invertebrates, especially ants, termites, spiders

IUCN STATUS
(1) EN = 5, VU = 1, NT = 2; percentage species in trouble = 9.5%; (2,3) percentage species in trouble = none

TOP | French Guianan Disc Frogs *(Synapturanus zombie)* have proboscis-like snouts, presumably for feeding on termites and ants.

ABOVE | The Neblina Frog *(Adelastes hylonomus)* is the sole species in the subfamily Adelastinae. It is confined to a small montane area of Venezuela, Guyana, and Brazil.

ASIAN NARROW-MOUTHED FROGS

RIGHT | The Painted Narrow-mouthed Toad (*Kaloula pulchra*) is a common species across tropical Asia and has been widely introduced elsewhere.

The largest of the three Asian subfamilies is Microhylinae, with 10 genera, 119 species, distributed from eastern Russia to Sri Lanka and south to Timor-Leste. The largest genus is *Microhyla* (51 spp.) containing numerous small, difficult to identify, leaf-litter inhabitants. *Kaloula* (19 spp.) contains the narrow-mouthed bullfrogs or pumpkin frogs which are short-headed and stout-bodied, including species such as the Banded Bullfrog (*Ka. pulchra*) which bears a pair of bold stripes that run along each side of the body and meet over the head.

Genus *Glyphoglossus* (10 spp.) contains frogs that are almost as round as balls, with short heads and relatively large forward-facing eyes. They are entirely fossorial, only emerging from their burrows to breed. *Metaphrynella* (2 spp.) are tiny, semiarboreal tree hole-dwelling species from Borneo (*Metaphrynella sundana*), and the Malaysian Peninsula (*Metaphrynella pollicaris*). Among the largest microhylines are the toad-like, globular frogs of the genus *Uperodon*, which occur from Pakistan to Sri Lanka and Myanmar.

DISTRIBUTION
(1) Tropical Asia; (2) Southeast Asia; (3) southern India

GENERA
(1) *Chaperina, Glyphoglossus, Kaloula, Metaphrynella, Microhyla, Micryletta, Mysticellus, Nanohyla, Phrynella, Uperodon;*
(2) *Kalophrynus;*
(3) *Melanobatrachus*

HABITATS
(1) Rainforest, swamps, and rice paddies; (2,3) Rainforest and swamps

SIZE
(1) ½ in (12 mm) ♂ Least Narrow-mouthed Frog (*N. perparva*) to 3 in (76 mm) Indian Globular Frog (*U. globulosus*); (2) ¾ in (18 mm) Robinson's Sticky Frog

(*Kalophrynus robinsoni*) to 2 in (50 mm) Black-spotted Sticky Frog (*K. pleurostigma*); (3) ½ in (9.7 mm) ♂ Galaxy Frog (*Melanobatrachus indicus*)

ACTIVITY
Nocturnal; terrestrial, fossorial, arboreal, secretive

REPRODUCTION
(1,2) Breeds in tree holes, pitcher plants, temporary

Key: (1) Microphylinae (red & blue),
(2) Kalophryninae (blue), (3) Melanobatrachinae (purple)

The Kalophryninae comprises just the genus *Kalophrynus* (27 spp.), the "sticky frogs" because they exude a sticky defensive residue when handled. They are small, short-legged, triangular frogs with sharply pointed snouts and a distinctive stripe running from the snout down the dorsolateral margin of the body. Distributed from China to Indonesia, they find their greatest diversity in Borneo (12 spp.). There is an endemic species on the Natuna Islands (*Ka. bunguranus*) in the South China Sea.

The monotypic Melanobatrachinae contains the Galaxy Frog (*Melanobatrachus indicus*) from the Western Ghats of southwest India, a "Lazarus species" rediscovered in 1997 after 100 years of "extinction." It is a stunning midnight blue/black frog covered with tiny white spots so that it resembles the night sky while its chest and hips are bright orange, this pigment extending forward to form rings around the forelimbs. The Galaxy Frog is Vulnerable, and one species each from the other two families are Critically Endangered—the Beilun Pygmy Frog (*Microhyla beilunensis*) and the Cameron Highlands Sticky Frog (*Kalophrynus yongi*).

ponds, mostly free swimming tadpoles; (3) Poorly known, believed direct breeder

DIET
(1,2,3) Forest floor invertebrates, i.e., ants, spiders; (1) also snails & small vertebrates (larger species)

IUCN STATUS
(1) CR = 1, EN = 7, VU = 8, NT = 7; percentage species in trouble = 20%; (2) CR = 1, EN = 2, VU = 1, NT = 1; percentage species in trouble = 22%; (3) VU = 1; percentage species in trouble = 100%

TOP | From India's Western Ghats, the Galaxy Frog (*Melanobatrachus indicus*) is a Vulnerable species.

ABOVE | The Black-spotted Sticky Frog (*Kalophrynus pleurostigma*) is the only *Kalophrynus* from the Philippines.

AFRO-MALAGASY
NARROW-MOUTHED FROGS

LEFT | Unlike its congenerics, the Maromandia Stump-toed Frog (*Stumpffia tetradactyla*) possesses only four toes.

OPPOSITE ABOVE | Madagascan Tomato Frogs (*Dyscophus antongilii*) are Near Threatened due to habitat loss and pollution but another factor is over-collection for the pet trade.

OPPOSITE MIDDLE | The Malagasy Rainfrog (*Scaphiophryne madagascariensis*) is a large frog from high elevations in eastern central Madagascar.

OPPOSITE BELOW | The West African Rubber Frog (*Phrynomantis microps*) is a distinctive savanna and woodland species.

Three microhylid subfamilies are endemic to Madagascar. The largest is Cophylinae with 8 genera and 115 species. It contains the genus *Mini*, which contains three species named to emphasize the fact they are among the smallest vertebrate species in the world: *Mini mum*, *Mini ature*, and *Mini scule*. At the other end of the scale are the burrowing frogs of the genera *Plethodontohyla* (11 spp.) and *Rhombophryne* (20 spp.). The Nosy Be Burrowing Frog (*R. testudo*) is almost as wide as it is long, while Boulenger's Digging Frog (*P. inguinalis*) may achieve 4 in (100 mm) SVL and feeds on scorpions.

Although microhylids are primarily terrestrial or fossorial in habit, Madagascar is home to several genera of arboreal microhylids, notably the Malagasy treefrogs, *Cophyla* (23 spp.), and the climbing frogs, *Anodonthyla* (12 spp.), the largest species being the Malagasy Giant Treefrog (*C. grandis*), females of which may achieve 3½ in (88 mm) SVL.

DISTRIBUTION
(1) Sub-Saharan Africa; (2) Tanzania; (3,4,5) Madagascar

GENERA
(1) *Phrynomantis*; (2) *Hoplophryne, Parhoplophryne*; (3) *Anilany, Anodonthyla, Cophyla, Madecassophryne, Mini, Plethodontohyla, Rhombophryne, Stumpffia*; (4) *Dyscophus*;

(5) *Paradoxophyla, Scaphiophryne*

HABITATS
Rainforest & dry woodland, also (1) dry or wet savanna, desert, and rocky inselbergs; (2) gardens (3) limestone forest (tsingy); (4,5) still or slow-moving watercourses

SIZE
(1) 1¼ in (30 mm) ♂ Marbled Rubber Frog (*Ph. annectens*) to 3¼ in (80 mm) ♀ Spotted Rubber Frog (*Ph. affinis*); (2) 1 in (23 mm) ♀ Usambara Black-banded Frog (*Pa. usambarica*) to 1¼ in (32 mm) ♀ Usambara Blue-bellied Frog (*H. rogersi*); (3) ½ in (9.7 mm) ♂ Manombo Tiny Litter Frog

Key: (1) Phrynomerinae (red & blue), (2) Holophryninae (blue), (3) Cophylinae (purple), (4) Dyscophiinae (purple), (5) Scaphophyninae (purple)

The Scaphiophryninae contains the web-footed frogs, *Paradoxophyla* (2 spp.) and Malagasy rainfrogs, *Scaphiophryne* (10 spp.). This last genus contains some true jewels, such as the multicolored Madagascar Rain Frog (*S. madagascariensis*) and the vivid-green Spiny Rainfrog (*S. spinosa*). The most famous Malagasy microhylids are the tomato frogs of the subfamily Dyscophinae and the genus *Dyscophus* (3 spp.), particularly the bright-red Tomato Frog (*D. antongilii*), which is popular in the pet trade. The Tomato Frog often inhabits disturbed secondary coastal forest, while the Sambava Tomato Frog (*D. guineti*) only inhabits mid-montane rainforest.

Sub-Saharan Africa is home to two microhylid subfamilies. The Phrynomerinae contains a single genus, *Phrynomantis* (6 spp.), the rubber frogs, which have rubbery skin. They are distributed across every African country from Senegal to Eswatini. Some are stunning, for example, the widely distributed red and black West African Rubber Frog (*Ph. microps*). The Rift Mountains of Tanzania are inhabited by the subfamily Hoplophryninae, containing the blue-bellied frogs (*Hoplophryne rogersi* and *H. uluguruensis*) and the Usambara Black-Banded Frog (*Parhoplophryne usambarica*). Ten Afro-Malagasy microhylids are Critically Endangered and 38 are Endangered.

(*Mi. mum*) to 4 in (100 mm) ♂ Giant Malagasy Burrowing Frog (*Pl. inguinalis*); (4) 2 in (50 mm) Antsouhy Tomato Frog (*D. insularis*) to 4 in (105 mm) ♀ Tomato Frog (*D. antongilii*); (5) ¾ in (17.5 mm) ♂ Masoala Web-footed Frog (*Pa. tiarano*) to 2½ in (60 mm) ♂ Rounded Rainfrog (*S. boribory*)

ACTIVITY
(1,3–5) Nocturnal; (2) Diurnal; all terrestrial, also (1,3) arboreal, fossorial; (4,5) semiaquatic

REPRODUCTION
Eggs laid in (1,4,5) ponds hatch, tadpoles free-swimming; (2) bamboo stems or leaf axils; tadpoles endotropic or exotrophic; (3) tree holes and

terrestrial nests, tadpoles endotrophic

DIET
(1) Specialized ant and termite predators; (2,3,5) forest floor invertebrates, including (3) scorpions (4) earthworms

IUCN STATUS
(1,4) Percentage species in trouble = none; (2) CR = 1, EN = 2; percentage species

in trouble = 100%; (3) CR = 9, EN = 35, VU = 5, NT = 3; percentage species in trouble = 45%; (5) EN = 1, VU = 2, NT = 2; percentage species in trouble = 42%

AUSTRALASIAN-ASIAN NARROW-MOUTHED FROGS

The final subfamily in the Microhylidae is the Asterophryninae, containing 17 genera and over 360 species, the bulk of which are distributed across the island of New Guinea (235 spp. in Papua New Guinea, 101 spp. in Indonesian New Guinea) and extreme northern Australia (24 spp.). The Asterophryninae is also represented in Indonesia (14 spp.), New Britain (1 spp.), Malaysian Borneo (1 spp.), the Philippines (2 spp.), and mainland Southeast Asia (6 spp.), most Asian microhylids belonging to the aforementioned Microhylinae.

The Asterophryninae is named for the genus *Asterophrys* (8 spp.), the New Guinea bush frogs. The genus is epitomized by the Blue-tongued Bush Frog (*A. turpicola*), a robust, widely distributed lowland species with a series of elevated tubercles that form a starry eyebrow over the eye, the origin of both the generic and subfamily names ("aster" = star).

LEFT | The New Guinea Bush Frog (*Asterophrys turpicola*) is a large frog with raised superciliaris around the eyes that resemble a starburst.

DISTRIBUTION
Australasia and Southeast Asia

GENERA
Aphantophryne, Asterophrys, Austrochaperina, Barygenys, Callulops, Choerophryne, Cophixalus, Copiula, Gastrophrynoides, Hylophorbus, Mantophryne, Oreophryne, Paedophryne, Siamophryne, Sphenophryne, Vietnamophryne, Xenorhina

HABITATS
Lowland and montane rainforest

SIZE
¼ in (8 mm) ♂ Amau Litter Frog (*P. amauensis*) to 2½ in (65 mm) New Guinea Bush Frog (*As. turpicola*)

ACTIVITY
Nocturnal; terrestrial, semi-fossorial

REPRODUCTION
Direct breeding in those species for which the reproductive strategy is known

Most asterophrynine frogs are small species, inhabiting leaf-litter or burrows. The largest genera are *Oreophryne* (71 spp.), the cross frogs, from New Guinea, eastern Indonesia, and the Philippines, and *Cophixalus* (70 spp.), the nursery or boulder frogs, from New Guinea, eastern Indonesia, and northern Australia. Both genera contain small frogs with rounded bodies, largish eyes, and toes with broad terminal discs. They are believed to be direct breeders, laying a small number of large eggs in the leaf-litter that hatch into froglets. Other genera that may be confused with *Oreophryne* and *Cophixalus* include *Austrochaperina* (29 spp.), the whistling or land frogs from New Guinea, northern Australia, and New Britain which is home to a single endemic, the New Britain Whistling Frog (*A. novaebritanniae*).

ABOVE | The male Papuan Egg-guarding Frog *(Oreophryne oviprotector)* remains with his eggs until they develop.

BELOW | The Northern Ornate Nursery Frog (*Cophixalus ornata*) inhabits the rainforests surrounding the tablelands around Cairns, Queensland, northern Australia.

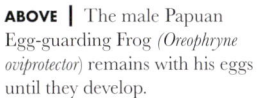

DIET
Presumed forest floor invertebrates

IUCN STATUS
CR = 13, EN = 8, VU = 15, NT = 7; percentage species in trouble = 12%

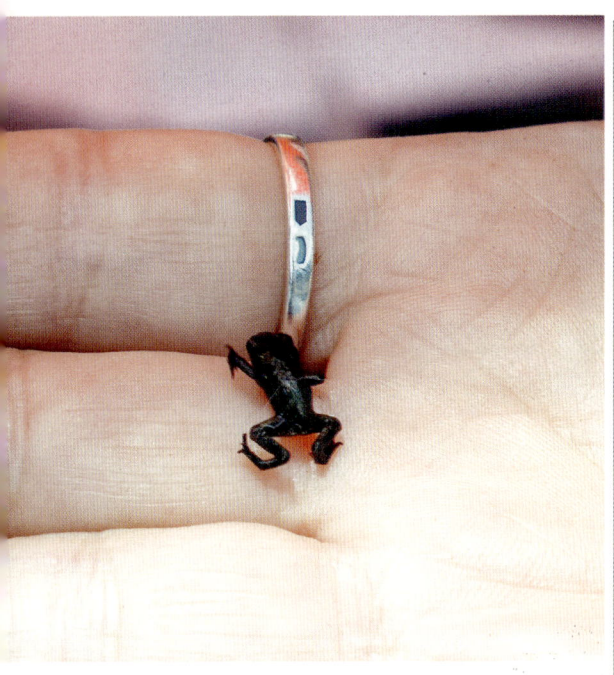

Also hiding in the New Guinea leaf-litter are the New Guinea narrow-mouthed frogs, *Callulops* (29 spp.); the large-eyed Mehely frogs, *Copiula* (15 spp.); the New Guinea wedge frogs, *Sphenophryne* (15 spp.); the Papuan ridge-nosed frogs, *Barygenys* (9 spp.); the high-elevation *Aphantophryne* (5 spp.), the Papuan pig-nosed frogs, *Choerophryne* (37 spp.), with their strange projecting snouts; and the New Guinea fanged frogs, *Xenorhina* (41 spp.), some of which bear a pair of vomeropalatine spikes in the roof of the mouth, for example, Zweifel's Fanged Frog (*X. zweifeli*). Some of these terrestrial frogs exhibit eye markings on their thighs that are exposed to intimidate a potential threat, a common tactic for frogs that merely hop and cannot leap to safety like long-legged ranids or treefrogs.

Most of the above genera contain short-legged species, but Asterophryninae also contains some longer-legged genera, for example the genus *Mantophryne* (5 spp.) from the Papuan Peninsula and the islands to its east, and *Hylophorbus* (12 spp.), which resemble small ranids.

Frogs in the Malagasy cophyline genus *Mini* (see page 228) are tiny but some of the New Guinea litter frogs belonging to the genus *Paedophryne* (7 spp.) are even more diminutive, with the male Amau Litter Frog (*P. amauensis*) achieving only ¼ in (8 mm) SVL, and likely claiming the title as the world's smallest vertebrate species.

Three asterophryine genera do not have their distributions centered over New Guinea. *Gastrophrynoides* contains two species of narrow-mouthed frogs from Borneo (*G. borneensis*) and the Malaysian Peninsula (*G. immaculatus*), mirroring the distribution of microhyline genus *Metaphrynella* (see page 226). They are poorly known in nature, with one rare specimen of the Bornean species being obtained from the stomach of a Giant River Frog (Discoglossidae: *Limnonectes leporinus*).

The genus *Vietnamophryne* (5 spp.) contains the Vietnamese dwarf frogs, four Vietnamese endemics and the Chiang Rai Dwarf Frog (*V. occidentalis*) from Myanmar and Thailand. The monotypic Tenasserim Cave Frog (*Siamophryne troglodytes*) inhabits a single cave in western Thailand, and is just one of 13 species of asterophrynine frogs listed as Critically Endangered, while a further 8 are Endangered, 15 Vulnerable, and 7 Near Threatened.

ABOVE | Arguably the world's smallest frog, the Amau Litter Frog (*Paedophryne amauensis*) from Papua New Guinea. Pictured is an adult.

OPPOSITE ABOVE | The Slender-snouted Frog (*Choerophryne gracilirostris*) inhabits the Müller Range, Papua New Guinea.

OPPOSITE | Slender Wedge-shaped Frogs (*Austrochaperina gracilipes*) occur both in Queensland and southern New Guinea.

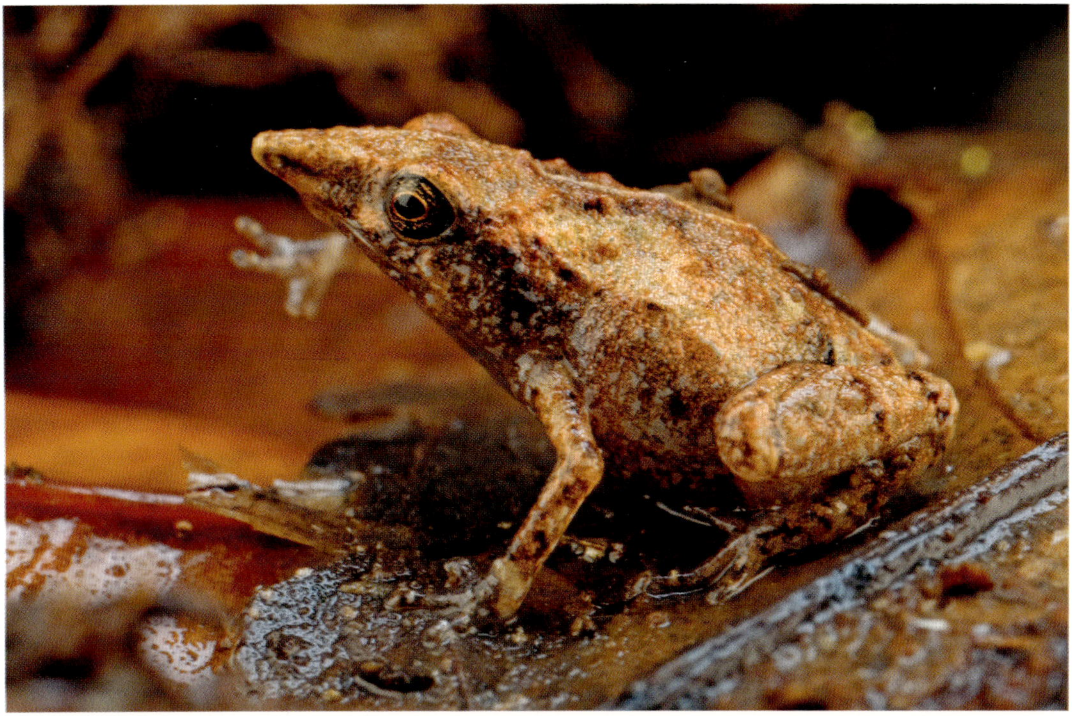

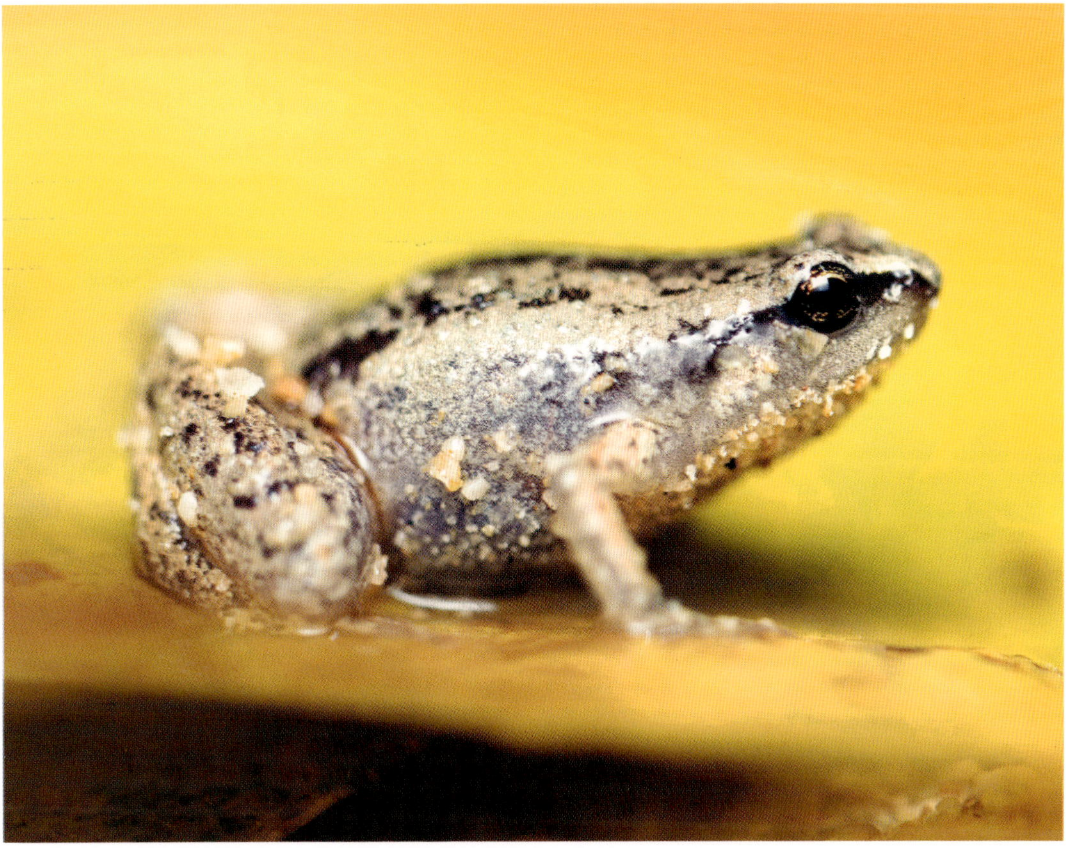

GLOSSARY

amphicoelous: the centrum of the vertebra is incompletely ossified around the notochord, a primitive condition today only seen in Ascaphidae, Leiopelmatidae, Scaphiopodidae, Pelodytidae, Pelobatidae, Megophrynidae, and Heleophrynidae (see also Opisthocoelous and Procoelous vertebrae).

amplexus: the position of a mating pair of frogs when the male grasps the female during mating. There are at least seven types of amplexus, the commonest being axillary (grasped behind the front legs) or inguinal (grasped around the waist).

anuran: a member of the amphibian order Anura which includes all frogs (and toads).

aposematism: warning colors, usually red, yellow, and black.

arciferal: a primitive condition in frogs where the epicoracoid cartilages of the pectoral girdle (supporting the forelimbs) are fused anteriorly but separate and overlapping posteriorly.

ASL: above sea level.

axillary: applied to amplexus when the male grasps the female behind the front legs.

Bidder's organ: an organ found in the gut of adult and larval true toads (Bufonidae) that appears to control sex hormones.

branchial basket: a cartilaginous structure that supports the gills of larval anurans.

buccal cavity: the oral cavity, the mouth.

caatinga: a northeastern Brazilian habitat comprising arid shrubland and thorn forest.

canthus rostralis: a line or ridge running from the front of the eye to the top of the snout, separating the dorsal and ventral surfaces of the head.

cavernicolous: cave-dwelling.

cephalic: of the head.

ceratohyal cartilage: largest cartilaginous component of the ventral hyoid arch.

cerrado: an eastern Brazilian habitat comprising tropical savanna.

chaparral: a western North American habitat comprising evergreen and deciduous woodland and scrubland.

chromatophore: cells in the skin that produce color.

cloaca: the common genital-excretory opening of amphibians, reptiles, birds, and monotreme mammals (differing from the single-purpose mammalian anus).

cloud forest: stunted forest found at high elevations which may be shrouded in clouds, exemplified by cool, wet, mist conditions, low tree canopies, and moss and lichen-covered branches, may also be called elfin forest or moss forest.

congeneric: member of the same genus.

conspecific: members of the same species

crypsis: camouflaged or exhibiting disruptive patterning that breaks up the outline.

deimatic: the presence of eye markings on the posterior body, the purpose being to intimidate a potential predator.

denticle: a small tooth-like structure.

dimorphism: when the sexes are morphologically different, such as the male exhibiting a crest or a larger head than a female of the same species.

diurnal: active by day.

dorsum: the upper surfaces of the body, dorsal (see also venter).

ectomorph: different morphologies of the same species that are influenced by its ecology in different locations or habitats.

ectothermic: the correct term for cold-blooded, an organism that controls its body temperature environmentally rather than metabolically, i.e., amphibians and reptiles as opposed to birds and mammals.

edge diffusion: a softening of the division between an anuran's color and the color of the background.

elfin forest: a type of cloud forest dominated by dwarf trees and small vertebrates.

endotrophic: a nonfeeding tadpole, often terrestrial in its ecology, that obtains its nutrition from the egg-yolk provided by the female (see also exotrophic).

epicoracoid: an element of the coracoid bone in the shoulder of a frog.

epipubis: a pair of forward-pointing bones attached to the pelvis, usually found in primitive mammals and protomammals but also present in New Zealand frogs (Leiopelmatidae).

epithelial: the external skin of the body including the skin of the gut.

estivate: a form of inactivity through a dry period.

exotrophic: a feeding tadpole, usually aquatic in its ecology, that feeds on other organisms.

firmisternal: an advanced condition in frogs where the epicoracoid cartilages of the pectoral girdle (supporting the forelimbs) are fused anteriorly and posteriorly.

fossorial: living life underground, a burrowing species.

fynbos: a South African habitat type comprising coastal scrub and heathland.

gallery forest: a type of forest found along rivers and other wetlands, usually through an open habitat such as grasslands or deserts, that acts as a linking corridor for genetic material from one forest block to another.

Gondwanaland: aka Gondwana, the great southern supercontinent that eventually fragmented into South America, Antarctica, Africa, Madagascar, the Seychelles, India, Australia, southern New Guinea, and New Zealand.

incertae sedis: meaning "of uncertain placement," a taxon that cannot with confidence be allocated to any known clade at the family or generic level.

inertial elongation: the catapult-like process by which certain anurans extend their tongues at speed to capture prey.

inguinal: applied to amplexus when the male grasps the female around the waist.

intromittent organ: an organ used by a male to transfer sperm to the inside of a female's genital tract, found in the tailed frogs (Ascaphidae).

iridophore: a chromatophore that reflects the colors green, blue, silver, and gold (see also melanophore and xanthophore).

IUCN: International Union for the Conservation of Nature, an NGO that determines the conservation status of animals and plants, using a series of abbreviations: EX = extinct, EW = extinct in the wild; CR = critically endangered; EN = endangered; VU = vulnerable; NR = near threatened; LC = low concern, and DD = data deficient.

lamella: structures under the terminal digit of the toe that enable a frog or gecko to walk up vertical surfaces.

lecithotrophic: all nutrition came from the egg-yolk provided by the female.

melanophore: a chromatophore that reflects the color black (see also iridophore and xanthophore).

meniscus: the surface meniscus is the surface of standing water which some invertebrates and amphibians are light enough to traverse without being immersed.

mesic forest: a forest growing in an area which receives regular, heavy rainfall, often in temperate regions.

metatarsal: the main bones of the toes.

monophyly: a monophyletic group contains an ancestor and all of its descendants, it is natural taxonomy that reflected evolution.

monotypic: a family with only one genus, or a genus with only one species.

MYA: million years ago, as in geological time.

myrmecophagous: feeding on ants or termites and their larvae and eggs.

Nobelian rod: cartilaginous rods that support the intromittent organ in the tailed frogs (Ascaphidae).

notochord: a flexible rod running between the gut and the vertebral column, the presence of which, at any stage of development, qualifies a taxon as a chordate.

nuptial pads: a pad that develops, usually on the thumb, of a male frog during the breeding season, and which serves as a secondary sexual characteristic to identify a male.

ontogenetic: a change that takes place when an organism matures from juvenile to adult, be it a color change or morphological change.

opisthocoelous: the centrum of the vertebra completely enclosed the notochord, allowing the vertebrae to move, but each vertebra is concave posteriorly and convex anteriorly, a primitive condition today only seen in Alytidae, Bombinatoridae, Pipidae, and Rhinophrynidae (see also Amphicoelous and Procoelous vertebrae).

orbitohyal muscle: part of the hyobranchial skeleton.

oviduct: the duct running from an ovary to the cloaca.

papilla (singular), papillae (plural): raised protuberances on the skin of an anuran.

parotoid gland: an enlarged gland behind the head on anurans, e.g., toads in the family Bufonidae, that stores and secretes noxious defensive toxins when the amphibian feels threatened.

patagium (pl. patagia): a flap or fold of skin that is used as a parachute by gliding frogs and lizards.

pectoral girdle: the bones of the shoulder that support the forelimbs.

perianthropic: living alongside man.

phragmosis: a behavior whereby a male frog blocks the entrance to a breeding chamber (a bromeliad) with his own ossified head to prevent other males accessing the eggs laid by the female.

phytotelma: a plant that contains water, such as a bromeliad or pitcher plant, which some anurans use as a nursery for their tadpoles.

presacral vertebrae: the vertebrae anterior to the sacrum or pelvis.

procoelous: the centrum of the vertebra completely encloses the notochord, so the vertebrae are jointed, but each vertebra is concave anteriorly and convex posteriorly, the modern condition seen in all modern frogs, the Neobatrachia, except Heleophrynidae (see also Amphicoelous and Opisthocoelous vertebrae).

puna: a type of grassland found in the high Andean valleys of Argentina and Chile.

saxicolous: a species that lives in rocky habitats.

scansorial: the ability to climb.

sclerophyll: a type of habitat which can sustain long periods of aridity with arid-adapted vegetation.

sexual dichromatism: when the sexes exhibit different colors or patterns.

sexual dimorphism: when the sexes exhibit different morphologies, from a head crest only seen in males, to a subtle difference in head sizes between the sexes.

sp. (singular), spp. (plural): an abbreviation used to indicate an unnamed single species or generalized species in a named genus, which is never italicized because it is not part of a scientific name.

spiracle: small posteriorly positioned holes on an anuran tadpole's head out through which the water exits having passed over the internal gills.

splash zone: an area above the high tide mark which is often splashed by the waves.

suctorial disc: a large disc surrounding the mouth of a tadpole with which it may anchor itself to a rock in torrential water to prevent it being washed away.

SVL: snout-to-vent length but excluding the length of the tail, as opposed to TTL which is total length including the body and the tail.

sympatry: where two more species live in the same geographical location.

synapomorphy: a character found in an ancestor and its descendants, therefore an ancient character which has been inherited.

synonymize: the process by which taxonomists sink one species into another having determined they are the same species, with the name retained being the older name by the Law of Priority.

thanatosis: a defensive tactic that involves playing death to avoid predation.

troglodyte: a cave dweller.

tsingy: a Malagasy name for a habitat comprising limestone karst with water eroded caves and fissures.

tympanum: the eardrum, seen externally in most frogs.

unkenreflex: a defensive tactic where the frog, such as a yellow-bellied toad (Bombatoridae), contorts and turns its body upside-down to display its brightly colored ventral surfaces in an effort to intimidate a potential predator.

urostyle: the fused lower bones of the spine located posterior to the sacrum (pelvis).

Valdivian temperate forest: a forest type found in southeastern Chile and Argentina which is dominated by ferns, bamboos, and evergreen deciduous trees.

venter: the lower surfaces of the body, ventral (see also dorsal).

vermiform: a wormlike shape, elongate.

viviparity: live-bearing, giving birth to near-fully formed neonates rather than laying eggs which contain only partially developed embryos.

vomerine teeth: a series of teeth or tooth-like structures located along the palette of the mouth.

vomeropalatine spikes: a pair of enlarged tooth-like processes in the roof of the mouth.

xanthophore: chromatophore that reflects the color yellow (see also iridophore and melanophore).

FURTHER READING AND USEFUL RESOURCES

(For combined amphibian and reptile guides and herpetological societies, see *Lizards of the World* (2021) or *Snakes of the World* (2023), both published by Princeton University Press)

BOOKS (GENERAL)

Crump, M. *In Search of the Golden Frog.*
University of Chicago Press, 2000.

Duellman, W.E. *Patterns of Distribution of Amphibians:*
A Global Perspective. John Hopkins University, 1999.

Gibbons, J.W. & M. Dorcas *Frogs: The Animal Answer Guide.*
John Hopkins University, 2011.

Halliday, T. *The Book of Frogs: A Life-size Guide to Six Hundred Species*
from Around the World. University of Chicago Press, 2016.

Heatwole, H. & J. Wilkinson *Amphibian Biology*
(multiple volumes) various publishers.

Moore, R. *In Search of Lost Frogs.* Bloomsbury Press. 2014.

Pough, F.H., R.M. Andrews, M.L. Crump, A.H. Savitsky,
K.D. Wells & M.C. Bradley. *Herpetology (4th edition).*
Sinauer Publishing, 2016.

Richardson, M. *Threatened and Recently Extinct Vertebrates of*
the World: A Biogeographical Approach. Cambridge University Press,
2023.

Stebbins, R.C. & N.W. Cohen *A Natural History of Amphibians.*
Princeton University Press, 1995.

Vitt, L.J. & J.P. Caldwell. *Herpetology: An Introductory Biology of*
Amphibians and Reptiles (4th edition). Academic Press, 2014.

NORTH AMERICA

Green, D.M., L.A. Weir, G.S. Casper & M.J. Lannoo
North American Amphibians: Distribution & Diversity. University of
California Press, 2013.

CENTRAL AND SOUTH AMERICA, AND THE WEST INDIES

Duellman, W.E. *Hylid Frogs of Middle America* (2 volumes).
SSAR, 2001.

EUROPE

Dufresnes, C. *Amphibians of Europe, North Africa & The Middle East:*
The Photographic Guide. Bloomsbury, 2019.

AFRICA AND MADAGASCAR

Channing, A. *Amphibians of Central and Southern Africa.*
Cornell University Press, 2001.

Channing, A. & M-O. Rödel *Field Guide to the Frogs & Other*
Amphibians of Africa. Struik, 2019.

Du Preez, L. & V. Carruthers *Frogs of Southern Africa:*
A Complete Guide (2nd edition). Struik, 2017.

Glaw, F. & M. Vences. *A Field Guide to the Amphibians and Reptiles of*
Madagascar (3rd edition). Vences & Glaw Verlag, 1994.

Henkel, F-W. & W. Schmidt. *The Amphibians and Reptiles of*
Madagascar, the Mascarenes, the Seychelles and the Comoros Islands.
Krieger Publishing, 2000.

ASIA AND ARABIA

Inger, R.F., R.B. Steubing, R.B., T.U. Grafe & J.M. Dehling.
A Field Guide to the Frogs of Borneo (3rd edition). Natural History
Books (Borneo), 2017.

AUSTRALASIA AND OCEANIA

Heatwole, H. & J.J.L. Rowley *Status of Conservation and Decline*
of Amphibians. CSIRO Publishing, 2018.

Menzies, J.I. *The Frogs of New Guinea and the Solomon Islands.*
Pensoft, 2006.

USEFUL WEBSITES

World Congress of Herpetology (WCH)
www.worldcongressofherpetology.org

Amphibians of the World database
https://amphibiansoftheworld.amnh.org

AmphibiaWeb database https://amphibiaweb.org

International Union for the Conservation of Nature (IUCN)
Red List of Threatened Species www.iucnredlist.org

Convention on International Trade in Endangered Species
of Fauna and Flora (CITES) www.cites.org

INDEX

PICTURE CREDITS

The publisher would like to thank the following for permission to reproduce copyright material: T = Top; B = Bottom; L = Left; R = Right; C = Center

Photograph courtesy of **Dr. Abdellah Bouazza**: 79BL. **Alamy Stock Photo** agefotostock: 49; Alf Jacob Nilsen: 21 (Row 6/C), 228; All Canada Photos: 21 (Row 5/L), 48, 61, 191; Anton Sorokin: 60, 83TR, 107TL, 120TL; Avalon Picture Library: 116; Biosphoto: 107CL, 118CR, 141B, 170, 172CL, 182; blickwinkel: 71, 85, 173, 181; Buddy Mays: 34–35; Buiten-Beeld: 64TL; Chris Mattison: 76, 131B, 149, 231BR; Dave Pinson: 95BL; David Hosking: 92; David Tipling Photo Library: 32TL; Dinal Samarasinghe: 205CL; Dorling Kindersley Ltd: 118TL; ephotocorp: 82, 203T; FLPA: 42–43, 80; George Grall: 24; Heather Rose: 145CR; Hemis: 100–101, 164, 225TR; imageBROKER: 21 (Row 6/R), 114CR, 215CL; Jessica Girven: 70TR; John Cancalosi: 19, 139T; John Sullivan: 83BR, 195TR; Kevin Schafer: 75; Luis Louro: 169CR; Marc Anderson: 27CL; Matthijs Kuijpers: 54, 68, 106, 165T; mauritius images GmbH: 13; Minden Pictures: 21 (Row 6/L), 21 (Row 7/R), 37, 44–45, 47, 86, 104, 129, 133CR, 156, 157TR, 175, 178BL, 183T, 199BR, 212, 214, 231TR, 233T; Morley Read: 39; Nature Picture Library: 29, 36, 53, 94, 96, 99TC, 111CL, 117CR, 121CR, 121BCR, 125TR, 126, 145TL, 153, 157BR, 162TCR, 168TR, 188, 198, 201T, 205T, 207CR; Papilio: 110; Premaphotos: 97CL; Robert Hamilton: 123; robertharding: 115; robin chittenden: 55; Sabena Jane Blackbird: 192; Sam Yue: 174CL; Science Photo Library: 33TR; Terence Waeland: 7; The Natural History Museum: 9; Tim Plowden: 21 (Row 7/C), 227CL; Universal Images Group North America LLC/DeAgostini: 10TL; WILDLIFE GmbH: 63, 229CR; Wirestock, Inc: 142; Zdeněk Malý: 171; Zoonar GmbH: 134. **Ardea.com** Micele Menegon: 213T. Photograph courtesy of **Dr. Axel Kwet**: 193T. Photograph courtesy of **Dr. Bikramjit Sinha**: 197. **Carlos Otávio Araujo Gussoni**: 109. **César Barrio-Amorós/Doc Frog Expeditions/CRWild**: 105TR. **D. Bruce Means**: 160. **David C. Blackburn**: 42BL. **Dreamstime** Alslutsky 21 (Row 2/R); Brandon Alms: 222; Chien Mu Hou: 174TL; Ecophoto: 189T; Farinoza: 144; Hotshotsworldwide: 67TL; Isselee: 150, 180; Jason Ondreicka: 135TR; Jason P Ross: 77T; Matthijs Kuijpers: 143CR, 151BL; Morley Read 21 (Row 2/L); Nathan Hutcherson: 70CL, 77BL; Nynke Van Holten: 103; Zdeněk Macát: 78. Photographs courtesy of **Professor Eli Greenbaum**, Ph.D.: 210–211, 217TR, 217BR. **Getty Images** Images from BarbAnna: 21 (Row 5/R), 223TL; Paul Starosta: 218, 221B; R. Andrew Odum: 51, 143BR. **Luke Verburgt**: 21 (Row 7/L), 89T, 220. **iStockphoto** alekseystemmer 21 (Row 3/C); Maria Ogrzewalska: 113CL. Photograph courtesy of **Dr. Mareike Petersen**: 219. **Marion Anstis** (From the book *Tadpoles and Frogs of Australia* by Marion Anstis, published by New Holland Publishers): 50. **Mark O'Shea**: 72, 91CR, 95T, 130TL, 130BL, 132, 146, 147B, 184, 195BR, 199TR, 230. **Martin Mandak**: 194. **Nature Picture Library**: Chien Lee: 67TR; Chien Lee/Minden: 22CL; Christian Ziegler: 12; Doug Wechsler: 23T; Emanuele Biggi: 52BL; Michael & Patricia Fogden: 32–33. Photo copyright **Rafe M. Brown**: 199TL. **Science Photo Library** Dante Fenolio: 74, 111T, 127, 152, 157CR; Lucas Bustamante/Nature Picture Library: 20; Dr. Morley Read: 168CL; Tom McHugh: 73. **Sean Michael Rovito**: 165BL. **Shutterstock** abcwildlife: 207TR; alex_gor: 232; Allen Lara Gonzalez 21 (Row 4/R); Beatrice Prezzemoli: 58; Chase D'animulls: 190; Cormac Price: 215T; Craig Cordier: 179, 187TL; David W. Leindecker: 133TR; Dirk Ercken 21 (Row 3/L); Dylan Leonard: 133BR, 187CR; EcoPrint: 178CR; Eric Isselee: 25, 41TR; Eugene Troskie: 210; Fabio Maffei 21 (Row 1/R); fivespots: 216; Henner Damke: 40TR; Holger Kirk: 226; IrinaK: 137CL; James A Somers 2; joaokloss: 113B; Kamil Strubar: 6, 21 (Row 5/C); Ken Griffiths: 52TR, 97TR; Klaus Ulrich Mueller: 119TL;

Krisda Ponchaipulltawee: 41CR, 229TR; Kristian Bell: 233B; Kurit afshen: 148–149; Luis Louro: 102CL; Manfredxy: 22; Manick,Jr 21 (Row 3/R); Marek R. Swsadzba: 66; mihirjoshi: 206; Mike Wilhelm: 137T; Milan Zygmunt: 141CL; Dr. Morley Read: 21 (Row 1/L), 114TL; NERYXCOM 21 (Row 1/C); Nynke van Holden: 56; Patrick K. Campbell: 143TR, 162BCR, 224; Petr Salinger: 18; prasanthdaskkm: 203BL; Ralfa Padantya: 177T; RealityImages: 201BR; reptiles4all: 39TL, 41BR, 193BR, 229BR; Rosa Jay: 84, 124, 138TL, 151T, 154; Rudmer Zwerver: 79T; sivananthan2001: 21 (Row 2/C), 177CR; Steve Byland: 136; Thorsten Spoelein: 38; Traxparent Wildlife: 30; Usha Roy: 155T; Valt Ahyppo 5; Vampflack: 169BR; Vision Wildlife: 147T; Vitalii Hulai: 130CL; zdenek_macat: 131TR, 176. **Smithsonian Institution** National Museum of Natural History: 225CR. **Twan Leenders**: 15, 221CR. **Tyrone Ping**: 25T, 89CR. **Václav Gvoždík**: 93, 211. **Wikimedia Commons** Arcasapos: 138BR; Benny Trapp: 65; Benny Trapp: 81; Bernard DUPONT: 189CR; Brian Gratwicke: 128CR; Brian Gratwicke: 140; Brian Gratwicke: 161; Brian Gratwicke: 166; Brian Gratwicke: 172TL; Coedilleradenahuelbuta: 98; © Darío De la Fuente: 155CR; Dariusz Kowalczyk: 31; David V. Raju: 21 (Row 4/L); David V. Raju; Davidvraju: 202; Davidvraju: 204; Davidvraju: 227TR; Diogo Luiz: 159; Gatomoteado: 139BL; Gionorossi: 122; H. Zell: 119TR; Hanyrol H. Ahmad Sah: 196; Hjv1986: 200; Holger Krisp: 117TR; Jalmirez: 112; © John Lyakurwa: 213BR; José Grau de Puerto Montt: 99TR; Marco Rada, Pedro Henrique Dos Santos Dias, José Luis Pérez-Gonzalez, Marvin Anganoy-Criollo, Luis Alberto Rueda-Solano, María Alejandra Pinto-E, Lilia Mejía Quintero, Fernando Vargas-Salinas, Taran Grant: 102CR; Marius Burger: 183BR; Michael F. Barej et al: 209; Mnapieceofnature: 185; Neil Birrell: 62; Nihaljabinedk: 90; Nihaljabinedk: 91; Nobu Tamra: 10TR; Oliver Angus: 88; Rafael M R Serra: 108; Renato Augusto Martins: 46; Renato Augusto Martins: 125BR; Renato Augusto Martins: 167; Ribeiro LF, Blackburn DC, Stanley EL, Pie MR, Bornschein MR: 158; Sandra Goutte, Jacobo Reyes-Velasco, Stephane Boissinot: 186; Santiago Ron from Quito, Ecuador Santiago R. Ron-FaunaWebEcuador: 120CL; Sergiocerrobravo: 163; USFWS Mountain-Prairie/Sara Armstrong: 128CL; Uzi Paz Pikiwiki Israel: 64CL; Xavier Heckmann: 105BL.

All reasonable efforts have been made to trace copyright holders and to obtain their permission for the use of copyright material. The publisher apologizes for any errors or omissions and will gratefully incorporate any corrections in future reprints if notified.

ACKNOWLEDGMENTS

The authors would like to thank all the herpologists, naturalists, and photographers who contributed images for this book as well as Dr. Rayna Bell. With special thanks to Maddy Fowler, Rebecca Morris, and Natasha Kruger for their review of the text. Thanks also to Joanna Bentley, Anna Southgate, Sara Harper, Dee Costello, Wayne Blades, and the rest of the Bright Press team.